MODERN CARTOGRAPHY
VOLUME ONE

GEOGRAPHIC INFORMATION SYSTEMS

The Microcomputer and Modern Cartography

Related Pergamon Titles of Interest

Books

HANLEY & MERRIAM (Editors)
Microcomputer Applications in Geology II

KOCH
Geological Problem Solving With Lotus 1-2-3

LISLE
Geological Structures and Maps

MALING
Measurements from Maps

Journals

Computers & Geosciences

Computer & Graphics

Ocean Engineering

Full details of all Pergamon publications/free specimen copy of any Pergamon journal available on request from your nearest Pergamon office.

Geographic Information Systems

The Microcomputer and Modern Cartography

Edited by

D. R. FRASER TAYLOR
Carleton University, Ottawa, Canada

PERGAMON PRESS
Member of Maxwell Macmillan Pergamon Publishing Corporation
OXFORD · NEW YORK · BEIJING · FRANKFURT
SÃO PAULO · SYDNEY · TOKYO · TORONTO

U.K.	Pergamon Press plc, Headington Hill Hall, Oxford OX3 0BW, England
U.S.A.	Pergamon Press Inc., Maxwell House, Fairview Park, Elmsford, New York 10523, U.S.A.
PEOPLE'S REPUBLIC OF CHINA	Pergamon Press, Room 4037, Qianmen Hotel, Beijing, People's Republic of China
FEDERAL REPUBLIC OF GERMANY	Pergamon Press GmbH. Hammerweg 6, D-6242 Kronberg, Fedeeral Republic of Germany
BRAZIL	Pergamon Editora Ltda, Rua Eça de Queiros, 346, CEP 04011, Paraiso, São Paulo, Brazil
AUSTRALIA	Pergamon Press Australia Pty Ltd., P.O. Box 544, Potts Point, N.S.W. 2011, Australia
JAPAN	Pergamon Press, 5th Floor, Matsuoka Central Building, 1-7-1 Nishishinjuku, Shinjuku-ku, Tokyo 160, Japan
CANADA	Pergamon Press Canada Ltd., Suite No. 271, 253 College Street, Toronto, Ontario, Canada M5T 1R5

First edition 1991

Library of Congress Cataloging-in-Publication Data

Geographic information systems: the microcomputer and modern cartography/edited by D. R. Fraser Taylor.
p.   cm.
1. Geographic information systems.   I. Taylor, D. R. F. (David Ruxton Fraser), 1937–
G70.2.G46   1991   910'.285—dc20   90-7762

British Library Cataloguing in Publication Data

Geographic information systems: the microcomputer and modern cartography.
1. Geography. Applications of microcomputer systems
I. Taylor, D. R. Fraser (David Ruxton Fraser) 1937–
910.285416

ISBN 0-08-040278-X Hardcover
ISBN 0-08-040277-1 Flexicover

Printed in Great Britain by B.P.C.C. Wheatons Ltd., Exeter

THIS BOOK IS DEDICATED TO THE MEMORY OF MY
FATHER, DAVID R. TAYLOR, WHO DIED IN MAY 1989

Preface

THE PURPOSE of this book is to consider the impact of two important technological changes on the discipline of cartography. The first is the technology of Geographic Information Systems which, in turn, has been made more feasible by the dramatic impact of the microcomputer revolution. Both of these have had a fundamental impact on cartography at all stages from data gathering to data presentation and use. These are considered in a systematic fashion in the chapters of this book.

Many individuals have helped in the preparation of this book, but special thanks must go to Eleanor Thomson and Barbara George for editorial and administrative assistance, and to Carleton University for providing research and administrative facilities without which the book could never have been written.

Ottawa, May 1990   D. R. FRASER TAYLOR

Contents

List of Figures

List of Tables

About the Authors

D. R. Fraser Taylor is Professor of Geography and International Affairs, Associate Dean (International) of Graduate Studies and Research and Director of Carleton International at Carleton University in Ottawa. He received his academic training at the Universities of Edinburgh and London in the United Kingdom. His research interests in cartography are mainly in the area of applied computer-assisted cartography, a field in which he has published widely. He also has a deep interest in Third World development problems. His most recent major book contributions are *The Computer in Contemporary Cartography*, Wiley, 1980; *Development from Above and Below, The Dialectics of Regional Planning in Developing Nations* (with W. B. Stohr), Wiley, 1981; *Graphic Communication and Design in Contemporary Cartography*, Wiley, 1983; and *Education and Training in Contemporary Cartography*, Wiley, 1985. Dr. Taylor has twice been elected President of the Canadian Cartographic Association. In 1984 he was elected Vice-President of the International Cartographic Association, and President in 1987. He was also elected President of the International Union of Surveying and Mapping in 1989.

Yves Brousseau received an M.A. in Geography from Laval University, Quebec City, in 1989. He is the first author of the micro-atlas *La Francophonie nord-américaine à la carte*, which was part of his M.A. thesis. He is currently working in Cameroon for a private firm engaged in natural resources development.

Barbara P. Buttenfield is an Assistant Professor and Director of the GIA Lab in the Department of Geography at the State University of New York at Buffalo where she teaches courses in computer and analytical cartography, cartographic design and map animation. She is a research scientist at the National Center for Geographic Information and Analysis (NCGIA). Her research interests include map generalization, scale-dependent geometry and development of knowledge base concepts for mapping. Dr. Buttenfield is a member of the American Congress on Surveying and Mapping (ACSM), the Canadian Cartographic Association (CCA), the North

Cartographic Information Society (NACIS) and the Association of American Geographers (AAG). She currently serves on the Board of Direction for the American Cartographic Association (ACA) and on the Science Policy Committee of the NCGIA. Her Ph.D. was awarded by the University of Washington in 1984.

Hinrich Claussen is an Engineering Scientist with the Mobile Communications Division of Robert Bosch GmbH in Hanover, West Germany. He was previously a Scientific Assistant in the Institute of Cartography at the University of Hanover where he received a Diploma in Geodesy in 1986. His technical education was in surveying and mapping in the Municipal Surveyor's Office in Bremerhaven.

David C. Coll is a Professor in the Department of Systems and Computer Engineering at Carleton University in Ottawa. He has taught and conducted research in the areas of communication theory, data and computer communications, continuous and discrete linear systems, digital signal processing, computers and digital systems, real-time computing and professional practice. His current research interests include digital image processing, image compressions and communications, broadband network applications and the philosophy of technology. In 1956 he became a Scientific Officer with the Canadian Defence Research Board where he worked on HF and meteor-burst communications and adaptive channel equalization, and in 1967 joined the faculty of Carleton University. Dr. Coll is Editor for CATV of the IEEE Transactions on Communications, Past Chairman of the Academic Requirements Committee of the Association of Professional Engineers of Ontario and President of DCC Informatics Incorporated, a company providing professional engineering services in information technology.

Claude Dufour received his M.A. in Geography from Laval University, Quebec City, in 1989. He was the main research associate in the preparation of the micro-atlas *Mines et minéraux à la carte* while completing his M.A. thesis which dealt with the design of a micro-GIS concerning the inventory and mapping of firms specializing in the recuperation and recycling of reusable goods. He is now working as a research associate in the preparation of an environmental atlas with computer-assisted cartography.

Stephen L. Egbert obtained a B.A. in Chinese from the University of Minnesota, his B.S. in Geography from Brigham Young University, and his M.A. in Geography from the University of Nebraska. He is currently completing work on a Ph.D. in Geography at the University of Kansas. His research interests include the design of interactive computer map displays and the application of geographic information systems to the analysis of landscape structure.

Timothy V. Evangelatos is Chief of Cartographic Development of the Canadian Hydrographic Service. He received a B.A. in Engineering Physics from McGill University, Montreal, and an M.Eng. in Systems Engineering from Carleton University, Ottawa. He worked in several engineering fields before becoming involved with the development of computers systems for hydrographic surveys and nautical charting, standards for the exchange of spatial data as well as the development of more effective systems for the compilation of nautical charts.

Jean-Philippe Grelot has received engineer's degrees from the *École Polytechnique*, Paris, in 1977, and *Corps des Ingénieurs Géographes* in 1979. He has been working at IGN-France in thematic cartography, becoming head of its Printing Division, and is now its Commercial Director. He also teaches automated cartography and the theory of cartography at the *École Nationale des Sciences Géographiques*. M. Grelot has participated in various activities of the International Cartographic Association, in particular its Commissions on Population Cartography, Thematic Cartography from Satellite Imagery, and Education, and Training. He is presently a Vice-President of the International Cartographic Association and of the French Cartographic Committee.

C. Peter Keller is a Professor in the Department of Geography at the University of Victoria, Canada, where he has taught since 1985. He has published on spatial analysis and geographic information systems. Professor Keller is actively involved in curriculum and facility development for computer cartography and geographical information systems. He has been Chairman of the Education Interest Speciality Group of the Canadian Cartographic Association.

Y. C. Lee is an Assistant Professor in Digital Mapping and Director of CanLab-INSPIRE, a laboratory in spatial sciences in the Department of Surveying Engineering, University of New Brunswick, Canada. He has a diploma in Cartography from ITC in The Netherlands, a B.Sc. on Computer Science from Simon Fraser University, an M.Sc. in Computer Science and a Ph.D. in Surveying Engineering from the University of New Brunswick. He has had many years of academic and industrial experience in various aspects of mapping before joining the UNB faculty, including the design and development of CARIS during 1980 to 1986. Besides teaching and research in digital mapping and geographic information systems, Dr. Lee is actively involved in consulting, and is also serving on several national and international committees in surveying and mapping.

David M. Mark is a Professor in the Department of Geography at the State University in New York at Buffalo where he teaches courses on algorithms and data structures for geographic information systems (GIS), on map use and cognitive science issues related to GIS. He is research scientist and

Chair of the Science Policy Committee at the National Center for Geographic Information and Analysis (NCGIA). His current research interests include development of user interfaces for GIS, algorithm design, and expert systems in cartography. Dr. Mark is a member of the American Congress on Surveying and Mapping (ACSM), the Canadian Cartographic Association (CCA) and the Association of American Geographers (AAG), where he is past Chair of the GIS Special Group. His Ph.D. was awarded by Simon Fraser University, Vancouver, in 1977.

Marc Miller is an Assistant Professor of Geography at Lavel University, Quebec. He received his B.A. (Hons.) in Geography from the University of Ottawa in 1975, and his M.A. in Geography at Laval University in 1987. He is currently involved in Ph.D. research concerned with electronic atlas and geographic information diffusion. His academic interests include computer assisted cartography, geographic information systems, quantitative methods and, more generally, the use of microcomputers in geography.

Donna J. Peuquet received a Ph.D. in Geography from the State University of New York at Buffalo in 1977. From 1978 to 1982 she was a research geographer with the U.S. Geological Survey in Reston, Virginia. In 1982 she joined the Geography Department at the University of California at Santa Barbara. She is currently an Associate Professor of Geography and an Associate of the Earth System Science Center at the Pennsylvania State University, where she has been since 1986. Her current areas of research interest include theory of spatial knowledge representation and spatial relationships, very large heterogeneous geographic databases, geometric algorithms and design of geographic information systems.

Jean-L. Raveneau received his *Doctorat de 3e cycle* in Geography from the University of Strasbourg in 1966. He is full Professor of Cartography in the Geography Department of Laval University, Quebec City, where he has been engaged in teaching and research in thematic cartography since 1963. He served as Department Chairman (1983–1986), Head of the Cartographic Laboratory (1970–1987) and successively as Editor and Managing Editor of the journal *Les Cahiers de géographie du Québec* (1970–1986). He is the author of articles in population and settlements mapping, the mapping of landscapes and the environment. He has contributed to the preparation of regional atlases, and is co-author of *L'interAtlas*, an instructional atlas published in 1986. His current interests are in the design and uses of computer mapping and electronic atlases for the communication of geographical information for instructional purposes. He is a member of the Canadian Cartographic Association.

Bengt Rystedt received his B.Sc. in Computer Sciences and his Ph.D. with a thesis on Computer Cartography from the University of Lund. From 1965 to 1972 he was a researcher in the Departments of Geography and Building

Function Analysis at the University of Lund, and from 1972 to 1984 he was responsible for the development of applications for urban and regional planning in the Swedish Land Data Bank System. He is now at the National Land Survey, and Head of the Information Systems Division in the Department of Research and Development. From 1987 and until 1991 he is Chairman of the International Cartographic Association National Atlas Commission.

Terry A. Slocum is Association Professor of Geography at the University of Kansas. He obtained his B.A. and M.A. degrees in Geography from the State University of New York at Albany, and his Ph.D. in Geography from the University of Kansas. He has published numerous articles in areas of experimental cartography and computer mapping. Currently he is completing a National Science Foundation grant dealing with the development and analysis of an information system for choropleth maps. His current research interests also include animated mapping and map display for GIS.

Nigel M. Waters is an Associate Professor in the Department of Geography at the University of Calgary, Canada, where he has taught since 1975. He has authored and co-authored various papers on computer mapping, geographic information systems, spatial decision support systems and knowledge acquisition. He currently writes a regular column in *The Operational Geographer* on these topics and is a contributing editor to *GIS World* where his "Edge Nodes" feature has appeared in a number of recent issues.

CHAPTER 1

Geographic Information Systems: The Microcomputer and Modern Cartography

D. R. FRASER TAYLOR

*Faculty of Graduate Studies and Research
and
Carleton International
Carleton University
Ottawa, Canada*

Introduction

Cartography is undergoing a period of rapid change as a result of the processes and products of the information revolution and there is a need to re-examine and redefine the nature of the discipline. Definitions are more than an exercise in semantics and are important because they focus attention on what a discipline sees as its major focus. This chapter will examine how cartography is changing and its relationships with the emerging technology of Geographic Information Systems. It will be argued that although technological change appears to be dominating the discipline the central issues for cartography, as it approaches the twenty-first century, are not primarily technological but conceptual. Existing concepts of the discipline, which include strong elements of neo-formalism and neo-positivism, are inadequate for modern cartography.

Cartography is an independent discipline. The first widely accepted formal definition of the discipline by the International Cartographic Association saw it both as a science and an art:

"The art, science and technology of making maps together with their study as scientific documents and works of art. In this context maps may be regarded as including all types of maps, charts and sections, three dimensional models and globes representing the Earth or any celestial body at any scale" (ICA 1973: 1).

1

Although published in 1973 the definition is, in fact, almost ten years older as it was based almost verbatim on the definition suggested by the British Cartographic Society (BCS) in 1964 (Board 1989) which read:

"The art, science and technology of making maps, together with their study as scientific documents and works of art. In this context, maps may be regarded as including all types of maps, plans, charts and sections, three-dimensional models and globes representing the earth or any heavenly body at any scale. In particular cartography is concerned with all stages of evaluation, compilation, design and draughting required to produce a new or revised document from all forms of basic data; it encompasses the study of maps, their historical evolution, methods of cartographic presentation and map use" (Board 1989: 1).

It is interesting that about the same time as BCS was offering its definition of cartography, the first experimental automated cartographic system was being demonstrated by David Bickmore at the Experimental Cartography Unit in London. From these early beginnings the influence of the computer in the discipline grew. The 1970s saw an increasing number of professional meetings and workshops devoted to the topic. The British Cartographic Society made automated cartography the theme of their annual meeting in 1973, and in December 1974 the American Congress on Surveying and Mapping held what was to be the first of an influential series of meetings on the computer in cartography, Auto-Carto I (ACSM 1976). Auto-Carto X is scheduled for 1991.

There was considerable debate over the significance of the computer to cartography as exemplified by the different views expressed in the chapters by Morrison and Rhind in *The Computer in Contemporary Cartography* (Taylor 1980) published in June 1980. Wolter had suggested that cartography was an "emerging discipline" (Wolter 1975) and Morrison argued that the computer was having a fundamental impact on the form and speed of that emergence. Rhind was more sceptical arguing that computer assisted cartography might be ". . . merely a new tool elevated by inflated academic claims to the level of a paradigm shift" (Rhind 1980: 25).

In parallel with the development of computer assisted cartography during the 1960s and 1970s was the development of the communication paradigm in the discipline. In theoretical terms it can be argued that no concept had greater theoretical influence over the decades of the 1960s and 1970s than communication. The concept of communication was, of course, not new to cartography but its formal acceptance as a key theoretical element in the discipline did not come until the late 1960s with major contributions being made by Board (1967), Bertin (1967), Kolácny (1969) and Ratajski (1973). The communication paradigm was in one sense a reaction to a traditional, formalistic view of cartography as being concerned only with techniques and technologies of map production, and for many cartographers communication came to represent the theoretical core of the discipline. By 1976 we find authors like Morrison arguing, "Cartographic scientists in many nations are now accepting this paradigm, and the impact of it on the discipline is becoming very pervasive. Cartography under this paradigm is a science" (Morrison 1976: 84). He went on to define cartography as "the detailed scientific study of a communication channel" (Morrison 1976: 96). Blakemore and Harley (1980), in a review of the history of cartography, said of the communication paradigm that "Today such concepts are assuming the status of an orthodoxy within cartography" (Blakemore and Harley 1980: 11), but even at the peak of its influence the communication paradigm did not receive universal acceptance. Many felt that the communication paradigm undervalued the cognitive aspects of cartography as a discipline, and this view was most forcibly expressed by Salitschev (1983). The 1970s saw an increasing degree of research concentration on cartographic communication, especially in North America, and fears were being expressed that the communication paradigm was dominating cartographic research to the exclusion of other issues.

Not all of this research proved to be useful to the design of better cartographic products and this contributed to a reassessment of the nature and direction of research into cartographic communication and design (Petchenik 1983; Olson 1983). The 1980s have seen a relative decline of the communication paradigm in cartography as interest in computer assisted cartography has increased.

In one sense the present interest in computer technology can be said to mark a return to a formalistic cartography dominated by questions of technology and techniques and some of the same criticism made of the communication paradigm can be applied to the emerging technological paradigm of the discipline. Salitchev, in criticizing the communication paradigm, argued that "Informatics, shunning the analysis of the content value and usefulness of scientific data, threatens the theory of cartography by turning into its Shakespearian Shylock, wrenching away its heart, the cartographic method of cognition. Without this, cartography is deprived of its creative functions and is brought down to the level of a technical means for the needs of other fields of knowledge" (Salitschev 1983: 31). To many computer scientists and GIS specialists the map is seen as a source of data, and cartography is seen only as a means of illustrating the output from their systems. To some cartographers the application of the computer is simply a means to automate existing map production with the aim of reducing costs and increasing the speed of production and revision. Neither cognition nor communication seem to enlighten these neoformalistic views of the discipline yet both of these are concepts vital to cartography. This argument will be considered in more detail later in this chapter.

The 1980s saw an accelerating pace of technological change as the computer found its way into all aspects of cartographic production. This was soon reflected in definitions of the discipline. Guptill and Starr described cartography as ". . . an information transfer process that is centred about a spatial data base which can be considered, in itself, a multifaceted model of geographic reality. Such a spatial data base then serves as the central core of an entire sequence of cartographic processes receiving various data inputs and dispersing various types of information products" (Guptill and Starr 1984: 1–3). Cartography, by this definition, is divorced from the map which is only one of a number of information products. For a discipline rooted in the map, this was a fundamental and radical change.

As cartography evolves a number of central questions emerge in seeking a new definition for the discipline (Board 1989), two of which are of particular significance. Can cartography be defined without using the term map, and does cartography include GIS or is it part of GIS? The International Cartographic Association established a Working Group on Cartographic Definitions in 1987 to study and revise the ICA definition of cartography and in 1989 the group reported on its work by suggesting two definitions for discussion. Cartography was defined as "The organization and communication of geographically related information in either graphic or digital form. It can include all stages from data acquisition to presentation and use." The map was not included in the proposed definition but it was defined independently as "A holistic representation and intellectual abstraction of geographical reality, intended to be communicated for a purpose or purposes, transferring relevant geographical data into an end-product which is visual, digital or tactile" (Board 1989: 4).

These suggested definitions have not yet been formally adopted and are likely to raise considerable debate. Although it is true that the map is only one product of a modern cartographic system, it is still central to cartography. As a complex, largely visual product, it should not be made trivial or ignored in defining cartography. In addition, the proposed definition by Board could equally well be applied to a Geographic Information System. An alternative definition of cartography is therefore suggested here reintroducing the map. Cartography is "The organization, presentation, communication and utilization of geo-information in graphic, digital or tactile form. It can include all stages from data preparation to end use in the creation of maps and related spatial information products." Under this definition the map and cartography continue to be inextricably linked although the importance of cartographic products in new forms is also recognized. The definition also requires that the map be defined and the definition suggested by Board, modifed semantically by Weiss (Board 1990, personal communication) is reasonable. "A representation or abstraction of geographic reality. A tool for presenting geographic information in a way that is visual, digital or tactile." A potential weakness in this definition of cartography is that it is difficult to define "related spatial information products" with precision.

The definition makes specific reference to products such as the map but this should not be read as excluding the consideration of *process* from cartography. Although cartography can be described as an applied, mainly visual, discipline, the study and understanding of process is important at all stages up to and including utilization. Indeed it can be argued that a failure to consider process in organization, presentation, communication and utilization will lead to an incomplete and inadequate set of products in general and inadequate maps in particular.

Cartography and Geographic Information Systems

There is no universally accepted definition of a geographic information system (GIS) and the terminology in the field includes a number of closely related terms such as land information system (LIS), land and resource information system (LRIS), urban information system (URIS), environmental information system (ERIS), and cadastral information system (CAIS). All of these deal with geographical and spatial data of various types and at various scales which seem to be the two major factors influencing the terminology used. Geographic information processing (GIP) would seem to be a more accurate term for this emerging field. The term Geomatics (Géomatique) has also gained some currency especially in Canada. However, the term GIS is now in such universal use that it will be difficult to replace it with an alternate term despite the fact that existing definitions of GIS are fuzzy and the topic of considerable debate.

To some GIS means only the suite of software used to analyse geographically referenced data. To others the term includes the hardware utilized by the system. Yet others would include all processes from data acquisition to data presentation. One recent definition reads: ". . . a system for capturing, storing, checking, integrating, manipulating, analysing and displaying data which are spatially referenced to the Earth. This is normally considered to include a spatially referenced computer database and appropriate applications software. A GIS contains the following major components: a data input sub-system, a data storage and retrieval sub-system, a data manipulation and analysis sub-system and a data reporting sub-system" (Stefanovic *et al.* 1989: 452).

Maps are central to all GISs both as a source of input data and as a means of demonstrating the output from such systems. There is general agreement that a GIS involves much more than automated map making and that a true GIS includes the ability to analyse data. There are at least two alternate views of the relationship between GIS and modern cartography. The first sees computer assisted cartography as part of a GIS (Tomlinson 1989; Blatchford and Rhind 1989). The alternate view is that GIS is a superstructure on a computer assisted cartography system (Stefanovic *et al.* 1989).

The difference between these two views is by no means a trivial one. At present it would be fair to say that the former is more prevalent than the latter. All GISs have a computer assisted cartography component but not all computer assisted cartography systems have GIS components. The situation is beginning to change as a new field of cartography, known as electronic mapping, continues to develop. Electronic mapping, as its name suggests, deals with maps and related cartographic products for use on electronic media. One such product is the electronic atlas. The electronic atlas is a new form of cartographic presentation and can be defined as an atlas developed for use primarily on electronic media. The electronic atlas is one product of an Electronic Mapping System (EMS) which is a system for the development and use of electronic maps. There is a major difference in emphasis between EMS and GIS. EMSs give major importance to the presentation and display of spatial information, and a good EMS should have all the functions of a GIS but in addition should have the capability of creating, storing and presenting material in a variety of different formats on electronic media (Siekierska and Taylor 1990). Electronic Mapping Systems will be a key to the future of cartography and offer new challenges and opportunities for cartographers.

The Convergence of Computer Assisted Cartography and GIS

Figure 1.1, which is based on one drawn by McLaughlin and Coleman (1989), provides a useful framework for discussion. The first point to be made is that both Automated Mapping Systems (shown on the left of the diagram) and the first generation GIS systems (shown on the right of the diagram) and in fact all subsequent developments have been dominated by advances in computer hardware and software. These were not developed primarily with cartography and GIS in mind and cartography is borrowing and adapting technologies developed for other purposes. There is, in a sense, a lack of control which is shared by many other disciplines affected by the rapid changes in the computer field. A significant difference between Fig. 1.1 and that originally drawn by McLaughlin and Coleman is the emphasis given to the central significance of technology. Hardware development came first and this continues to be the case. Software was then developed to utilize the hardware, and finally uses were sought to which this technology could be applied. Many authors have pointed out that this is a complete inversion of the desirable sequence. Developments in hardware continue to outpace developments in software and applications and there is very limited evidence that this is changing in a substantial way.

Early developments in Computer Assisted Cartography and GIS shared these developments in hardware and software but tended to utilize them in different ways. First generation Data Base Management Systems, for example, tended to be vector oriented in the case of GIS and raster oriented in the case of Automated Mapping Systems.

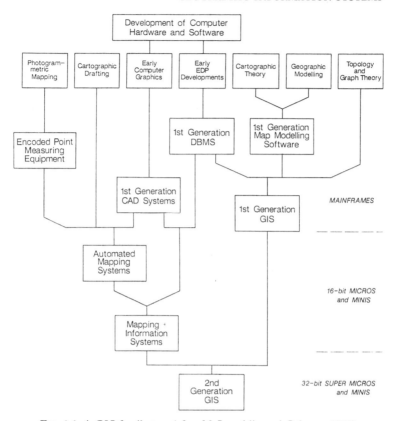

FIG. 1.1. A GIS family tree (after McLaughlin and Coleman 1989).

Many cartographers used the comuter initially to automate existing map production. The terminology used reflects this approach which was described as "automated mapping". The procedures, processes and data structures often followed closely the existing approaches they were designed to replace, and the "litmus test" used for the products of such systems was that they were indistinguishable in form, content and standards from those produced by existing conventional methods. Automated mapping was raster based and there were early relationships with electronic processing developments in photogrammetry as well as remote sensing and geodesy. Little attention was paid in these developments to geographic information processing other than in terms of what was required to manipulate the data to produce the map. The emphasis of these early systems was often on the production of the topographic base map, and the underlying cartographic approach was clearly a formalistic one.

Somewhat independently of this stream was another set of developments shown on Fig. 1.1 which led to the emergence of the first generation of GIS systems. As the diagram shows, the provenance of these developments lay in geographic modelling, topology and graph theory and some cartographic theory, especially from geographic and thematic cartography. The emphasis here was largely on the manipulation of spatial data. The approach used was vector based. Often the map was seen primarily as a data input source or as a display device for the system. In most instances the maps displayed were of poor design and quality as cartographers were rarely involved in designing these systems. Little thought seems to have been given to the need to effectively display and communicate the information that these systems could generate and few systems were capable of producing acceptable map graphics. Although cognition was an important concept, cartographic cognition was rarely predominant in the systems.

Parallel to the development of Automated Mapping and GIS were early developments in computer graphics which led to the first generation of Computer Aided Design Systems. These have continued to develop, especially for engineering purposes.

Each of these three sets of developments was somewhat independent of the others although there was a common major driving force in each case— the computer technology itself. Developments in each field were heavily dependent on the hardware and software available, little of which, as mentioned earlier, was developed specifically for the uses under discussion here. In addition many of the first generation GIS products could be described as solutions in search of a problem and solutions which often were neither technologically, administratively nor financially particularly effective. The International Geographical Union (IGU), for example, has estimated that as many as 65% of first generation GIS systems failed.

The three streams described on the diagram, Automated Mapping, Computer Graphics and GIS are now beginning to converge and second generation GIS systems together with Electronic Mapping Systems are vast improvements on the earlier systems. A major reason for this is again technological change.

The Impact of Microcomputer Technology

The early developments in automated cartography, computer graphics and GIS were built on expensive mainframe technology. Towards the end of the 1970s the first microcomputers appeared in the market place and microcomputer technology exploded during the 1980s. In 1986 the first micro utilizing the revolutionary Intel 8386 chip, the Compaq Desk Pro, was introduced followed closely by a range of other products using this chip, such as the IBM Personal Two series. Bryden (1989) has likened the introduction of the 8386 chip to the revolutionary changes which took place in 1989/90 in the Soviet Union and Eastern Europe in terms of the impact this had. It was a quantum leap in microcomputer technology of almost equal significance to the introduction of the microcomputer itself. Towards the end of 1989 an even more powerful chip, the Intel 8486, was introduced, and in January 1990 IBM released the most powerful of its Personal System/2 family, model 70486, based on this chip. The impact has been dramatic. Computer power previously only available on expensive mainframes is now available at rapidly decreasing costs on machines which are cheap, reliable and readily available. Both personal computer and workstation technology continue to develop at a rapid rate.

This has been accompanied by equally rapid developments in data storage technologies such as optical disks, CD-ROM, WORM, optical tape (Baker and Barkman 1990) and related products together with vastly improved telecommunications technologies such as ISDN, fibre optics and satellite transmission systems. These are more fully described in the chapters by Coll and Lee.

A third development has been the rapid emergence of new display technologies such as HDTV, Liquid Crystal Devices and high resolution colour monitors described in the chapter by Slocum and Egbert. The development of the laser printers now available at reasonable cost has put the production of high quality graphics of publication quality in the hands of a vastly increased number of people. Further development of colour laser printers will enhance this capability enormously.

These technological developments have facilitated the convergence of various streams of development and, as is shown on Fig. 1.1, previous "solitudes" are now beginning to interact in a much more positive way and as a result cartography is poised to contribute to the information era in new and exciting ways.

New Directions for Cartography in the Information Era

It can be argued that the 1990s will be the "Digital Decade" and that the already substantial impact of digital technologies on cartography will continue to accelerate. Despite the importance of issues such as data capture, data base structure, software, hardware and standards which are all covered in subsequent chapters of this book, the central issues which should determine new directions for cartography are not primarily technical. Indeed, it can be argued that if the prevailing technological paradigm is allowed to predominate then much will be lost. To allow our understanding and definition of cartography to be determined by a technological imperative would be a mistake.

In 1984 the term "A New Cartography" was introduced (Taylor 1985) to draw attention to the dramatic impact of information technologies on the discipline. Part of the concluding sentence of the argument made in 1985 read, "The greatest . . . challenge facing cartography does not lie in teaching or learning new techniques, but in creating a radically new concept for our discipline" (Taylor 1985: 22).

It can be argued that this challenge remains today and, if anything, is more pressing. Ramirez has argued that, "Unfortunately, the advent of new technologies such as computer-aided mapping, land information and geographic information systems (before the theoretical understanding of cartography was completed) decreased the interest and amount of research in theoretical cartography. As a result of that, there is no widely accepted 'cartographic theory' yet that can be used as the basis for the cartographic process" (Ramirez 1990: 19). The conceptual and theoretical development of cartography as a discipline has been retarded, if not diverted, by the explosion of interest in automated mapping and GIS. The neo-formalist and neo-positivist paradigms which have emerged as a result are somewhat sterile and too many modern cartographers are primarily technological specialists with a limited understanding of the problems to which cartography can be usefully applied and with a very limited and unimaginative view of the discipline.

GIS is a technique: Cartography is a discipline, and by definition cartography must be more than an element in a Geographic Information System. In the convergence between GIS and Computer Assisted Cartography, described earlier in this chapter, it is important not to lose sight of the importance of cartographic process.

Although cartography is an applied discipline there is a need to develop and maintain an intellectual non-applied core and this is largely lacking in the discipline.

Waters, drawing on the work of Bloom and Roszak and others, draws attention to some of the key issues relating to the debate in geography over GIS which he calls ". . . the new stampede towards geography's 'Holy Grail'" (Waters 1989: 32). He points out that GIS is decidedly positivist and as a result there is little dialogue with either the humanists or the realist/structuralist paradigms in geography.

Similar arguments can be applied to cartography. Brian Harley, for example, has argued that maps are neither scientific nor objective and that the notion of cartography as a progressive science is a "myth partly created by cartographers in the course of their own professional development" (Harley 1989: 2). Harley, utilizing the "post-modern" ideas of Foucault and Derrida, attempts to redefine the nature of maps as representations of power, echoing in a more subtle way some of the earlier ideas of the controversial French geographer Lacoste. To Harley, cartography should not be understood from the perspective of the dominant epistemology of scientific positivism but should be rooted in social theory. Maps are seen as discourses or texts and their metaphysical and rhetorical nature must be explored. Considerable importance is attached to Foucault's notion of power and the need to look at the social and political dimensions of cartography—how the map works in society as a form of power-knowledge and how it is used in this respect. Cartography must be seen in its social context, which by definition is culturally specific and changes over space and time.

Cartographers, in their quest for scientifically "objective" products, should not lose sight of the fact that maps have been and continue to be made for a variety of purposes and can never be truly objective because one of their distinguishing features "... is that they focus attention *selectively* on regions of space, features, objects and themes" (Visualingam 1989: 27). Even as new cartographic products and processes emerge this is unlikely to change. The map is an *abstraction* of reality: it is not reality itself. Nor is communication purely objective: it is a *rhetorical* process. At issue is just how explicit the rhetoric is.

Art seems to have been driven out of contemporary cartography because, as Arthur Robinson has observed, "... what can be explained to a computer is science and everything else is art" (Robinson 1989: 95). It can be argued, however, that if modern cartography is to progress then the need is not only for more science but also for more art. Waters, quoting Roszak, draws attention to the 'crucial importance of theoretical imagination, hypothesis, speculation and inspired guesswork" (Waters 1989: 31).

Expert systems, artificial intelligence and computerization require a quantification of cartography but following Roszak as paraphrased by Waters, there is perhaps just as much need for quality and for consideration of criteria such as relevance, coherence and insight which is sadly lacking in modern cartography. As Waters has observed, "The information in the system must be illuminated by ideas and concepts" (Waters 1989: 31), and by questions such as whose interests are served by the products that cartographers create. Such questions make many cartographers uncomfortable and are rarely posed explicitly in the cartographic literature.

There are three concepts which could inform and improve the technological formalism of modern cartography. These are cognition, visualization and communication, and together these might provide a strong conceptual and theoretical basis for the discipline.

Cognition, Communication and Visualization

Cognition and communication are not new concepts for cartography but they have taken a new significance in the information era. Previously a major problem for cartographers was obtaining sufficient information to map. The information revolution has seen an explosion of data and has opened up a whole new range of possibilities of topics which can be mapped. The need to convert data into useful information has never been greater and the map and related cartographic spatial information products are ideal media for the organization, presentation, communication and utilization of the growing volume of information which is becoming available.

Maps have always answered the question "where", but in the information era they must also answer a variety of other questions such as "why", "when" and "by whom", and must convey to the user an understanding of a much wider variety of topics than was previously the case.

Cognition of reality has always been an objective of cartography and as Papp Vary (1989) has pointed out it is difficult to separate the form (cartographic representation) from the content (the representation of reality).

The demands for the understanding of the complexities of modern society are great and there are few disciplines better placed to respond to these demands than cartography. The map has always been a means of navigation, but can take on fundamental importance in helping "navigate" through an increasingly turbulent sea of new data and information on a wide variety of topics not previously considered of central importance to cartography. Cartography must supplement and complement its topographic and locational products with thematic products which will increase our understanding of the world in which we live.

The concept of a map allows the relationships between a wide variety of both qualitative and quantitative data to be organized, analyzed, presented, communicated and used in a way which no other product can match. As Arthur Robinson and Barbara Petchenik point out in a fundamental and seminal study on maps, *The Nature of Maps* (Robinson and Petchenik 1976), the map is as old as human history and is present in all societies. The concept of maps and mapping, whether mental constructs or physical products is so basic that psychologists often talk of mapping strategies as central to the understanding of how the human brain functions.

Cartographic cognition is a unique process as it involves the use of the human brain to recognize patterns and relationships in their spatial context. This is difficult to replicate by the software currently available to most GIS systems which is constrained to some extent by the nature of vector based data structures which are often topological, sequential or object oriented. As Cartwright has pointed out, "In many cases the full information content of data may be realized only by displaying it in a spatial context or mapping the data" (Cartwright 1989: 9). The advent of GIS has improved cartographic cognition very substantially and some aspects of the process have been quantified, but there still remains an important intuitive element and the process is not one which is fully understood.

Cartographic communication also takes on a new importance in the information era and new challenges present themselves. These involve both the creation of new products to improve the effectiveness of the transmission of information and a better understanding of the process of communication. Many maps and information products derived from them will be in quite different forms from the traditional paper product. The human brain's perception of these new electronic images is quite differnt from that of the paper products. Research in both cognitive and human factors psychology will be of interest to cartographers. Relatively little cartographic research has been carried out in this area, but its importance is great and should lead to a revitalization of research and applications in the field of cartographic communication. This was a major conclusion of the volume on *Graphic Communication and Design in Contemporary Cartography* (Taylor 1983), but progress has been slow and is reviewed by Slocum and Egbert in this volume.

The new technologies allow interesting and innovative relationships between cognition and communication. The emerging field of visualization is a good example of this. Visualization is a field of computer graphics (McCormick *et al.* 1987) which is exploring both the *analytical* and *communication* power of visual interpretation. Robertson has reviewed some of the implications of visualization for cartography (Robertson 1988). "Visual representations of data aim to exploit effectively the ability of the human visual system to recognize spatial structure and patterns. This can provide the key to critical and comprehensive applications of the data benefiting subsequent analysis, processing or decision making" (Robertson 1988: 243). Visualization attempts to provoke intuitive appreciation of the salient characteristics of a data set and to "Map relevant aspects of the data, which may or may not be visual in nature on to visual representations that can be understood easily and intuitively by the observer" (Robertson 1988: 243).

Visualization research suggests that if images similar to the natural three dimensional world are used as a model, then both analysis and communication may be improved. "It is becoming increasingly recognized that effective utilization of the visual systems' spatial analyzing capabilities can depend on exploiting, and not confounding, its natural sense processing mechanisms. This suggests using natural sense properties whose attributes can be appreciated distinctively and intuitively, as a vehicle for representing data variables with similar attributes of interest" (Robertson 1988: 251). For cartographers, this new technique is an exciting extension of methods for the imaginative presentation of data which have been present in cartography since earliest times. Visualization is of course dependent upon new computer techniques of analysis and display of data and is therefore precise, exact and accurate. Although scientific in nature its effective use requires the theoretical imagination called for by authors such as Roszak. It could provide an interface which would increase the dialogue between the positivists, humanists and realists/structuralists which Waters (1989) argued is missing from the current scene. Visualization is a scientific tool but it demands artistry, imagination and intuition in its application. Figure 1.2 gives a visual impression of the relationships which are basic to cartography and which are being combined and inter-related in new ways in the information era. If cartography is to progress, then equal emphasis must be given to all three sides of the triangle shown. At present the interest in new

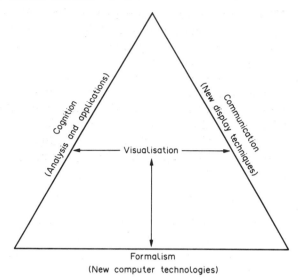

FIG. 1.2. A conceptual basis for cartography.

computer technologies is receiving an inordinate share of attention. These new technologies are of course of great importance and cartographers must give them that attention but not to the exclusion of cognition and communication. Visualization is an interesting example of how all three aspects can be combined. Visualization requires the use of the latest computer technology, whereas at the same time it offers a visual method of understanding complex relationships while communicating to the viewer the reality of the topics considered in new ways on the screen.

Earlier reference was made to the fact that the computer is driving the art out of cartography. It is perhaps ironic that computer graphics techniques such as visualization may in fact bring a renewed artistic and imaginative element back into the discipline, albeit one based on a strong scientific and technological base.

Beyond GIS: Electronic Atlases and Multi-Media Systems

Although GIS technology is over twenty-five years old, it is only just beginning to realize its potential in the market place. Tomlinson (1989) has argued that 1988/89 marked an important cross-over point for the technology as it truly came of age in a commercial sense.

Two major newsletters have been established, *GIS World* and *GIS Forum*, and there were at least fifteen major symposia on GIS in 1989 in North America alone drawing at least 20,000 participants. In a comprehensive survey of the GIS market which retails for about $1500 Daratech

Incorporated (1989) provides information on more than 160 GIS vendors and products and the company argues that "The market for geographic information systems (GIS) has the potential to become one of the most dynamic computer systems-related businesses of the 1990s." GIS is being applied to an increasing range of topics such as environmental monitoring, natural resource management and site planning, public management and planning, facilities management and mineral extraction, but the market is beginning to change, ". . . the nature of the market for GIS is about to change dramatically. As social and technological developments accelerate and more and more organizations in private industry become aware that GIS can help them understand and manage their business information more effectively, the need for geo-related technology will increase substantially and industry growth will far exceed its current 32% rate. Already GIS is being used for business information applications such as demographic market analysis, delivery service dispatching, and property management and the future points to the proliferation of GIS technology-based business and data processing systems" (Daratech Inc. 1989: 3).

There will continue to be a relationship between cartography and GIS but there are already signs that cartographers are finding GIS somewhat passé, and are moving to new approaches which incorporate GIS as only one of a number of useful technologies in the creation of new products based on or derived from the electronic map.

The electronic atlas is one such product (Siekierska and Taylor 1990). There are several systems and products already on the market and these can be classified according to the extent of their analytical capabilities (i.e. GIS), together with the level of interaction they allow. At one end of the spectrum are products which can be described as "view only" systems, which simply allow the display of data. More complex systems permit increasing degrees of dynamic interaction and analysis.

The Domesday Project (Openshaw and Mounsey 1987) is one of the better known electronic mapping systems and a key component of this system is the ordnance survey map base of Britain at various scales, which is used as one means of accessing a very extensive database stored on two optical disks. Although the user has considerable flexibility in terms of choice of scale, Domesday has limited analytical capabilities and is largely a static database.

Some of the individuals who participated in the Domesday Project are also involved in the development of CORINE, an electronic mapping system on the environment being developed for the European Economic Community (Mounsey and Briggs 1988).

Several electronic atlases are being developed using HyperCard and Apple Computers has produced two somewhat limited products called HyperAtlas and MacAtlas. Hypertext, on which HyperCard is based, was developed in 1945, but the current software is much more recent.

HyperCard is especially useful for cartography as it allows the non-sequential association of data elements and has been used by Raveneau and his colleagues at Laval University to produce some innovative products, more fully described in a subsequent chapter of this book.

One of the first electronic atlases was the Digital Atlas of the World produced by Delorme Mapping Systems in 1986 (Taylor 1987), designed to operate on an IBM compatible PC, as was the Electronic Atlas of Arkansas (Smith 1987). A PC Atlas of Sweden (Arnberg and Justusson 1989) with extensive analytical capabilities is under development, and there are several other electronic atlases and databases in the development stage.

The creation of the Digital Chart of the World (DCW) at the scale of 1:1 million by the Defense Mapping Agency of the United States of America is under way, and this important product will be available by 1992. The digital database is being designed to allow the user to perform a number of analytical functions, and the DCW will become an important building block for a variety of electronic mapping products.

Electronic atlases and electronic mapping systems are being complemented and supplemented by multi-media systems which utilize not only visualization techniques but a variety of other approaches such as video and sound. Desktop video may well become as significant in the 1990s as desktop publishing was in the 1980s, and the linkage between microcomputers, compact disk and video is now well advanced with the first multi-media systems hardware likely to be available for sale in 1991. The major technological challenge is the compression of video data in a form which will allow efficient storage on CDs, and the major stumbling block in this process is not so much the technology as reaching agreement on which video-compression standard is to be used.

Although still very much in the experimental stage, some interesting work on multi-media systems for cartography has already been completed. Louise Guay (1990) demonstrated this in the context of the Electronic Atlas of Canada. The Multi-Media Atlas may well be the wave of an exciting future for cartography, and the concepts of such an atlas ". . . involve visualization of information, schematization, comparative analysis, ordering, animation, dynamic modelling, projection, random navigation, Hypertext, data bases, and a capacity for processing and interactivity" (Guay 1990: 2).

Like visualization, multi-media electronic mapping systems wil involve all three conceptual elements shown on Fig. 1.2. They depend upon sophisticated computer technologies but are at the junction of these new technologies with both cartographic cognition and cartographic communication. To the visual has been added the use of other senses such as hearing and eventually all of the other senses, including touch and smell, may be involved.

Louise Guay comments, ". . . we need images to read the world, to understand and define it. An atlas cannot be other than electronic and multimedia if it is to represent the world today, since in our world machines record images: They read, interpret and employ images. There is no other way of representing our country and using the data our machines generate than by using the language of modern science . . . . Today's scientific imagery systems make the invisible visible and provide a multiplex vision of the world" (Guay 1990: 6).

In the demonstration given by Guay and her colleagues, the electronic map was used as the central organizing tool for a multi-media presentation of the export of wheat from Canada.

The core of the demonstration was an electronic map of the route and volume of wheat exports through the Great Lakes beginning at Thunder Bay at the lake head. The electronic map is interactive, allowing the user to call up video and textual material at any point along the route. Thus the viewer can use a video of a ship being loaded with wheat at the lake head and follow that ship on its journey all the way through the Great Lakes and the St. Lawrence Seaway system to the ocean. Textual or visual material can be accessed at any time the user wishes. When the ship reaches a major lock system, for example, images of the ship passing through the locks together with statistical and other data can be easily called up.

As technology advances, the capability of multi-media electronic mapping systems will be expanded. To a full range of forms of analyzing and presenting alphabetic and numeric data together with images and sounds will be added elements such as the ability to use senses such as smell. In dealing with pollution, for example, the systems may well be able to give the user an idea of the noxious odours present in many highly polluted locations.

Multi-media systems are of course not confined to cartography and, in fact, as so often is the case with new computer based technologies, are being developed primarily for other purposes such as entertainment, general education and training. The Peugeot automobile manufacturing company, for example, is using a multi-media system in training its mechanics, and the major market for multi-media systems may well be home entertainment combining video and Compact Disk in new ways.

The cartographic potential of such systems is, however, great. The map can be part of the database of these new systems, but at the same time can be an important tool for organizing the information which such systems contain. Louise Guay again captures this well.

"Maps used to be the primary instruments for navigation, exploration and discovery. Now they have become the tools for interactive computer navigation. These models of the world have been transformed into worlds of models. A map whose system incorporates the architecture of space includes not only its organization but also our way of using that space and of representing and simulating it. In other words, we will be navigating through knowledge. Maps have given us a superb and dynamic way of learning.

"Multi-media systems employ cartography, maps and graphs, along with multiple interfaces between media and the notions of legends and translations interwoven throughout. The metaphors of travel and geographic exploration pervade multi-media systems since we are setting out on new continents of multi-sensory languages.

"At first the world appeared to us in either-or auditory and written, hence visual forms. Now it will be transmitted and communicated by systems more closely resembling human communication in that all the senses are brought to bear" (Guay 1990: 2).

Conclusion: Cartography as an Applied Discipline

This chapter has argued that there is a need for a revitalized conceptual base for cartography, but in the final analysis cartography will be judged not by its concepts and theories but by the value which society places on its products. The need for cartography's traditional product—the paper map—will continue, but this market is unlikely to grow at a rapid rate, and if cartography is to flourish then the paper map will have to be supplemented and complemented by new products, and the topics to which cartography is applied must be expanded. This will require imagination and initiative by cartographers. This is primarily a human not a technological problem.

References

American Congress on Surveys and Mapping (ACSM) (1976) *Proceedings of Auto Carto 1*, Washington.

Arnberg, W. and B. Justasson (1989) "Concepts of the PC-version of the Atlas of Sweden", Stockholm.

Baker, R. and K. Barkman (1990) "One terabyte per reel optical tape recording and its application to remote sensing data storage", *Technical Papers 1990 ACSM-ASPRS Annual Convention*, Volume 4, Image Processing/Remote Sensing, Washington, pp. 43–53.

Bertin, J. (1967) *Semiologie Graphique*, Gauthier-Villars, Paris.

Blakemore, M. J. and J. B. Harley (1980) "Concepts in the history of cartography: a review and perspective", *Cartographica*, Vol. 17, No. 4, pp. 1–120.

Blatchford, R. and D. W. Rhind (1989), "The ideal mapping package" in Rhind, D. W. and D. R. F. Taylor (eds.) *Cartography Past, Present and Future*, Elsevier, London.

Bloom, A. (1987) *The Changing of the American Mind*, Simon & Shuster, New York.

Board, C. (1967) "Maps as models" in Chorley, R J. and P. Hagget (eds.) *Models in Geography*, Methuen, London.

Board, C. (1989) "Report to ICA Executive Committee for the Period 1987–89", Working Group on Cartographic Definitions, Budapest.

Bryden, R. (1989) "GIS: an industry perspective", Workshop on Strategic Directions for Canada's Surveying, Mapping, Remote Sensing and GIS Activities, Energy, Mines and Resources Canada, Ottawa.

Cartwright, J. C. (1989) "Map maker for the MacIntosh", *GIS World*, Vol. 2, No. 6, p. 9.

Daratech INc. (1989) *Geographic Information Systems, Markets and Opportunities*, Cambridge, (Mass).

Guptill, S. and L. E. Starr (1984) "The future of cartography in the information age", *Computer Assisted Cartography Research and Development Report*, International Cartographic Association, Washington.

Guay, Louise (1990) "A multimedia atlas", National Atlas Information Services Opportunities Seminar, Energy, Mines and Resources Canada, Ottawa.

Harley, B. (1989) "Deconstructing the map", *Cartographica*, Vol. 26, No. 2, pp. 1–20.

ICA (1973) *Multilingual Dictionary of Technical Terms in Cartography*, Steiner, Weisga Den.

Kolácny, A. (1969) "Cartographic information—a fundamental term in modern cartography", *Cartographic Journal*, Vol. 6, pp. 47–49.

McCormick, B. H., T. A. Defanti and M. D. Brown (1987) "Visualisation in scientific computing", *Computer Graphics*, Vol. 21, No. 6.

McGlaughlin, J. D. and D. J. Coleman (1989) "Land information management into the 1990s", United Nations Economic and Social Council, E/Conf. 81/BPI, New York.

Morrison, J. L. (1976) "The science of cartography and its essential processes", *International Yearbook of Cartography*, Vol. 16, pp. 84–87.

Morrison, J. L. (1980) "Computer technology and cartographic change" in Taylor, D. R. F. (ed.) *The Computer in Contemporary Cartography*, Wiley, Chichester.

Mounsey, H. and D. Briggs (1988) "CORINE: an environmental data base for the European community", *Proceedings Third International Symposium on Spatial Data Handling*, Sydney, Australia, pp. 387–406.

Olson, J. (1983) "Future research directions in cartographic communication and design" in Taylor, D. R. F. (ed.) *Graphic Communication and Design in Contemporary Cartography*, Wiley, Chichester.

Openshaw, S. and H. Mounsey (1987) "Geographic information systems and the BBC Domesday interactive videodisk", *International Journal of Geographic Information Systems*, Vol. 1, No. 2, pp. 173–179.

Papp Vary, A. (1989) "The science of cartography" in Rhind, D. W. and D. R. F. Taylor (eds.) *Cartography Past, Present and Future*, Elsevier, London.

Petchenik, B. B. (1983) "A map maker's perspective on map design research 1950–1980" in Taylor, D. R. F. (ed.) *Graphic Communication and Design in Contemporary Cartography*, Wiley, Chichester.

Ramirez, J. R. (1990) "Implementation of the cartographic language for the universal computer aided mapping systems", *Technical Papers 1990 ACSM-ASPRS Annual Convention*, Vol. 2, ACSM/ASPRS, Washington, pp. 19–29.

Ratajski, L. (1973) "The research structure of theoretical cartography", *International Yearbook of Cartography*, Vol. 13, pp. 217–228.

Rhind, D. W. (1980) "The nature of computer assisted cartography" in Taylor, D. R. F. (ed.) *The Computer in Contemporary Cartography*, Wiley, Chichester.

Robertson, P. K. (1988) "Choosing data representations for the effective visualisation of spatial data", *Proceedings Third International Symposium on Spatial Data Handling*, Sydney, Australia, pp. 243–252.

Robinson, A. H. (1989) "Cartography as art", *Cartography Past, Present and Future*, Rhind, D. W. and D. R. F. Taylor (eds.) Elsevier, London.

Robinson, A. H. and B. B. Petchenik (1976) *The Nature of Maps: Essays Towards Understanding Maps and Mapping*, University of Chicago Press, Chicago.

Roszak, T. (1986) *The Cult of Information*, Parthenon, New York.

Salitschev, K. (1983) "Cartographic communication: a theoretical survey" in Taylor, D. R. F. (ed.) *Graphic Communication and Design in Contemporary Cartography*, Wiley, Chichester.

Smith, R. (1987) "Electronic atlas of Arkansas: Design and Operational Considerations", *Proceedings of the 13th International Cartographic Conference*, Morelia, Mexico, pp. 161–166.

Stefavonic, P., J. Drummond and J. D. Muller (1989) "ITC's response to the need for training in CAL and GIS", *INCA International Seminar Proceedings*, Dehra Dun, pp. 450–460.

Siekierska, E. M. and D. R. F. Taylor (1990) "Electronic mapping and electronic atlases: new cartographic products for the information era—the electronic atlas of Canada", in Kartographenkongress Wien 1989 (Wiener Schriften zur Geographie und Kartographie, Bd. 4), Wien.

Taylor, D. R. F. (ed.) (1980) *The Computer in Contemporary Cartography*, Wiley, Chichester.

Taylor, D. R. F. (ed.) (1983) *Graphic Communication and Design in Contemporary Cartography*, Wiley, Chichester.

20   D. R. FRASER TAYLOR

Taylor, D. R. F. (ed.) (1985) *Education and Training in Contemporary Cartography*, Wiley, Chichester.

Taylor, D. R. F. (1987) "New methods and technologies in cartography: the Digital World atlas", *International Yearbook of Cartography*, pp. 219–224.

Tomlinson, R. F. (1989) "Presidential Address: Geographic information systems and geographers in the 1990s", *The Canadian Geographic*, Vol. 33, No. 4, pp. 290–298.

Waters, N. (1989) "Big bytes, micro bytes, Tid bytes and nibbles: Do you sincerely want to be a GISD analyst?", *The Operational Geographer*, Vol. 7, No. 4, pp. 30–36.

Wolter, J. (1975) "Cartography, an emerging discipline", *The Canadian Cartographer*, Vol. 22, No. 4, pp. 19–37.

Visualingam, M. (1989) "Cartography, maps and GIS in perspective", *The Cartographic Journal*, Vol. 1. 26, No. 1, pp. 26–32.

CHAPTER 2

Cartographic Data Capture and Storage

Y. C. LEE

Department of Surveying Engineering
University of New Brunswick
Fredericton, New Brunswick, Canada, E3B 5A3

Introduction

Cartographers acquire data from diverse sources to compile a map. For topographic maps these include air photographs, data from remote sensors, field notes, coordinate lists and existing maps. Thematic mapping and GIS rely on an even wider range of data sources such as census reports, meteorological records and historical documents. In modern cartography the variety of data sources is increasing, resulting in an exponential growth in the volume of data serving cartographic applications and GIS.

High quality maps require considerable storage space because long strings of digits are required to store coordinates of high resolution, and many points spaced not more than 0.1 mm apart at map scale are needed to depict smooth lines. Quick sampling of a low density 1 : 50,000 Canadian National Topographic Series map of gentle terrain shows that the total length of lines forming contours and hydrography is about 90 metres at map scale. The map measures 78.5×54 cm covering a ground area of 39×27 km. With a spacing of 0.1 mm, about 9×10^5 points in two dimensions (or 1.8×10^6 coordinates) are required to represent these lines.

Several new data capture devices have emerged in recent years, many of which generate data in digital form. The impact of some of these devices on modern surveying and mapping has been investigated by Paiva (1990), Tiller (1990), Crossfield (1989), Dale and McLaughlin (1988), and many others. Regarding the hardware and software developments in GIS, a review is given by Lee and Zhang (1989). This paper surveys some of the new data capture devices and recent developments in storage technology.

The data capture devices and systems reviewed in this chapter include total stations, satellite positioning systems, photogrammetric workstations, remote sensors, line digitizers and raster scanners. They all generate positional data with various degrees of resolution and accuracy. Total stations integrate three devices together into a unit providing the functions of angle measurement, distance measurement and digital data storage. They help eliminate a great deal of manual data recording. Satellite positioning systems reduce high precision surveying to an efficient one-man operation. Photogrammetric workstations allow map compilation from aerial photographs with the help of computers. Remote sensors provide extremely current data for thematic and topographic mapping at small scales. Line digitizers are, up to the present day, the mainstream instrument for converting existing maps into digital form, and raster scanners are automated digitizers gradually gaining popularity. In the review of storage media, we will concentrate on optical disks.

Many of the cartographic data capture devices complement each other. For example, Hartl (1989) reviewed the complementary roles of remote sensing and satellite navigation. Faig et al. (1990) and Oimoen (1987) described the application of digitizers in analytical photogrammetry. Konecny et al. (1987) discussed the use of satellite imageries on analytical photogrammetric instruments. Several researchers (e.g. Faig and Shih 1989; Kinlyside 1988) have investigated the combination of satellite positioning systems with photogrammetry.

Map making often involves several data capture devices, each affecting the accuracy of the map in various degrees. We will leave the discussion on the accuracy of each method to later sections, and will introduce here map accuracy requirements to provide the background for comparison. In-depth treatments of cartographic errors and map accuracy can be found in Maling (1989) and Muller (1987).

Different map accuracy standards are used by mapping organizations around the world. The Canadian (Moore 1989; EMR 1976), American (Rhind and Clark 1988; Slama 1980), and NATO (NATO 1970) standards for top quality topographic maps specify a planimetric accuracy of about 0.5 mm at the publication scale, and a vertical accuracy of half the contour interval. The planimetric accuracy corresponds to approximately 12.5 metres on the ground for a 1 : 25,000 map. Controls necessary to generate high quality maps are much more accurate than this, usually in the order of centimetres for national geodetic control networks (Maling 1989).

In the rest of this chapter, we will first describe each of the data capture devices and systems in more detail. This is followed by a review of the developments in computer storage, particularly that of optical disks.

Data Capture Devices and Systems

Total Stations

Traditionally, detail ground surveying is performed using a theodolite, a distance measuring device, and a field book. A total station integrates these three items into one electronic surveying instrument. It is the combination of an electronic theodolite and an electromagnetic distance measuring (EDM) device, and is capable of recording and transmitting readings in computer-compatible form to a data collector (Fig. 2.1). The data collector is a small hand-held unit which is itself programmable, provides storage for hundreds and even thousands of measured points, and can connect to external storage such as disk or tape. A survey of total stations on the market is given by McDonnell (1989), and of data collectors by McDonnell (1987). Munjy *et al.* (1989) review the characteristics of software for total station survey systems.

Compared with other methods such as satellite positioning, surveying with total stations is relatively labour intensive because it normally involves two persons. The need for the two to communicate limits the distance between survey stations to 500 m. However, the high precision of this instrument allows the survey network, consisting of many stations, to cover several kilometres.

Total stations provide three-dimensional measurements in real time with extremely high precision. For angular measurements, precision of total stations ranges from one second to one minute of arc. Instruments for common detail mapping have a precision of 5 seconds which is the angle subtended by 2.4 cm at 1 km. The EDM device, either built into a total station or provided as an attachable accessory, allows the measurement of distance with a typical precision of about 5 parts per million (ppm) which is equivalent to 5 mm in 1 km. With such high precision, total stations are suitable for the most demanding projects such as surveying the national geodetic networks.

The use of total stations improves efficiency and accuracy by eliminating human recording errors and allowing the digital manipulation of survey data. A study by Paiva (1990) shows that compared to conventional methods, modern survey systems using total stations can offer 40% improvement in time and 21% improvement in cost.

Total stations usually form part of a survey system which consists of a laptop computer in the site office and probably a workstation in the home office (Fig. 2.1). The purpose of the laptop computer is to process the data accumulated by the data collector, provide some editing, data reduction, and quality control functions in the site office. Data recorded include vertical and horizontal angles, distances, instrument and target heights, temperature and pressure, and some descriptive data such as point numbers. Sometimes the data collector is bypassed and data is loaded directly into the laptop computer. The home office workstation provides programs to perform other post-survey processing such as least squares adjustments and digital terrain modelling.

The design and accuracy of total stations will continue to evolve. It can be expected, however, that the data collector will remain as the most diversified component in terms of functionality. From the early versions which are primarily for data collection, models that provide considerable computing power have emerged. Developments in computing technology will make the data collectors easier to use and their built-in programs more sophisticated.

Satellite Positioning Systems

Satellite positioning systems are all-weather systems offering centimetre accuracy. Unlike total stations, surveying using satellite positioning systems does not require stations to be intervisible and can be performed by one person.

The American system is called the Global Positioning System (GPS). A similar system called GLONASS [GLObal NAvigation Satellite System) will be implemented by the Soviet Union. Analysis by Kielland and Casey (1990) projected that dollar savings from the use of GPS technology over conventional hydrographic surveying methods could average 48%, while productivity increase could average 42%. Applications of GPS are described in Wells and Kleusberg (1990) and Henderson and Quirion (1988); a survey of GPS equipment on the market is given by Reilly (1989).

An in-depth explanation of the principles of GPS can be found in Wells (1986) and Wells *et al.* (1982). The following is a brief introduction. If the three-dimensional positions, $P_1 (X_1, Y_1, Z_1), P_2 (X_2, Y_2, Z_2), P_3 (X_3, Y_3, Z_3)$, of three control stations are known, and their distances $(R_1, R_2, \text{and } R_3)$ to an unknown station are given, it is possible to calculate the three-dimensional position of the unknown station (Fig. 2.2). In GPS the control stations are satellites whose positions in orbit can be determined accurately.

To allow distance measurement, the satellites carry extremely accurate atomic clocks which typically will lose or gain an average of only one second in 30,000 years. Each satellite continuously transmits time and its position to earth. If a clock of the same accuracy is maintained by the receiver at the unknown ground station, its distance to a satellite can be computed simply by noting the time it takes for the signal to reach ground.

Using this method, the three-dimensional position of the ground station carrying an accurate clock can be determined if three satellites are visible simultaneously. However, the clock used by the receiver on the ground is normally less accurate. Because the two clocks are slightly out of synchronization, introducing a clock bias, the calculated distance to a satellite is not exact and is therefore called the pseudo-range. The clock bias is a systematic error which can be determined if information from a fourth

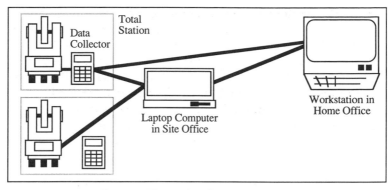

FIG. 2.1. The total station survey system.

satellite can also be obtained. That is, knowing P_i (three-dimensional positions) and R_i (pseudo-ranges) from four satellites, we can uniquely solve for the four unknowns which are the position (X, Y, Z) of the ground station and the clock bias which will eventually yield the true clock time for the receiver (Fig. 2.2).

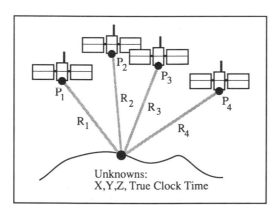

FIG. 2.2. GPS positioning system.

In order to provide continuous four satellite coverage of the earth, 21 primary satellites plus three active spares (sometimes denoted as $21 + 3$ satellites) will orbit the earth by mid 1990 (Green et al. 1909). During early 1990, 12 satellites are in operation while four experimental ones are still in orbit although they are no longer functional. The 24 GPS satellites will follow 12-hour orbits at a height of 10,924 nautical miles (about 21,000 km) in six orbital planes. The satellites will be controlled by a master control station (MCS) at Falcon Air Force Base in Colorado Springs, Colorado, in the heart of continental United States, and the MCS is supported by monitor stations around the world.

The Russian GLONASS will operate at a similar altitude, use also $21 + 3$ satellites, and will have an orbital period of 11 hours and 15 minutes. The system will be in full operation between 1991 and 1995. To some observers, a similar system from a rivaling nation reduces the risk of intentional destruction of the GPS satellites (Burgess 1990).

A more urgent concern to the military is the enemy's access to GPS positioning signals. GPS allows the quick determination of absolute position with an accuracy of 20 metres horizontally and 30 metres vertically. Signals to allow such computation are restricted only to the U.S. military and authorized users. The general public, and hence also the enemy, can determine position with an accuracy of about 40 metres horizontally and 60 metres vertically. A position fix can usually be obtained within one second. The "selective availability" policy of the U.S. Department of Defense will further degrade this accuracy to about 100 metres horizontally and 150 metres vertically. This is done by deliberately introducing errors in the "civilian band" of the GPS satellite signals.

In contrast with absolute techniques which yield position directly, differential techniques exploit the characteristic that the deliberate and other GPS errors on two nearby receivers are similar and hence can be made to cancel each other out. This differential mode of GPS survey returns a baseline vector (X, Y, Z) between two stations, and can yield much more accurate measurements, typically 2 to 5 metres, in distance. As a consequence, absolute position with accuracy up to 2 metres can also be obtained using differential methods if the coordinates of one of the stations is known accurately.

Very high accuracy of about one part per million (ppm) can be obtained using the slower static methods which require measurements to be carried out for several hours at a station. Recent research attempts to find kinematic methods (Ashkenazi and Summerfield 1989; Logan 1988; Remondi 1985) that could produce comparable accuracy but require much less measurement time per station.

Photogrammetric Workstations

Photogrammetry determines three-dimensional positions from a stereo pair of aerial photographs usually at a scale smaller than the maps they produce. For 1 : 25,000 maps, photographs at a scale of 1 : 40,000 are commonly used (Ghosh 1987). These photographs can be obtained from wide angle aerial cameras with 150 mm focal length at a height of 6 km. Because of the high resolution of black and white photographic films, conventionally 50 lines per millimetre (2540 lines per inch) (Slama 1980) but reaching 125 lines per millimetre (6350 lines per inch) with very high resolution aerial films

(Becker 1988), the photographs can be enlarged many times to yield high precision in measurement. The enlargement is normally done through the optics of the photogrammetric instrument.

The key concept in photogrammetry is parallax which is the apparent displacement in the image of an observed object due to different points of view. The amount of parallax is a function of the distance between the object and the observer. If the observer is vertically above ground, then parallax measurements can yield the height of ground features. In order to make the measurement of parallax more convenient, stereoscopic pairs of photographs of the ground to be surveyed are taken using cameras mounted on aeroplanes.

The use of aerial photographs for mapping allows parallax measurements to be done in the office. Photogrammetric stereoplotters enable an operator to see the terrain in stereoscopic view (called a stereo model) as one would from above ground. An important step before parallax measurements can be made is to ensure that the operator obtains an identical view of the terrain as seen by the camera in the plane when the photographs are taken. The reconstruction of the camera position and attitude (the rotations of the camera with respect to the ground) is called orientation.

There is generally a lack of accurate data for the camera position and its attitude although recent research shows that it is possible to determine them using kinematic GPS techniques (Faig and Shih 1989; Kinlyside 1988) to the accuracy required by photogrammetry. A more conventional approach is to use control points on the ground with known coordinates to help orientation. Control points can be obtained by identifying conspicuous features or creating visible marks on the ground and carrying out a field survey to determine their positions. They can also be obtained through a photogrammatric technique called aerial triangulation. Photogrammetric control points usually require an accuracy of 0.1 mm at map scale in horizontal position and 0.1 of the contour interval in height. For 1 : 25,000 mapping, horizontal controls with an accuracy of about 2.5 metres are required. Most of the maps at that scale have a contour interval of 10 metres (Ghoch 1987) implying that vertical controls should be accurate to about one metre.

There are hence two important tasks performed by a photogrammetric plotter: orientation and position measurement in three dimensions. Traditional analogue instruments use mechanical linkages to orient the stereo model. A high quality analogue photogrammetric plotter, called a precision plotter, provides a resolution of about 10 microns (10×10^{-3} mm) and an accuracy of 15 microns at the image scale.

The analytical photogrammetric plotter (Fig. 2.3) was conceived by U. V. Helava at the National Research Council (NRC) of Canada (Helava 1963; Helava 1957). It keeps the stereo model oriented digitally and thereby eliminates many of the mechanical parts. A high quality analytical instrument typically provides a resolution of 1 micron and an accuracy of 3 microns. It is also possible to drive the analytical plotter from the host computer so that the operator can locate a point on the stereo model with known ground coordinates, or can program the optics to follow a predefined path. This greatly facilitates measurements along a profile.

An analytical plotter, combined with a graphics monitor and a host computer, form the components of a photogrammetric workstation (Fig. 2.3). The host computer is where the computer mapping programmes reside, the analytical plotter is the three-dimensional digitizing station, and the graphics monitor shows the digitizing result. Sometimes, the graphics monitor serves also as the interface for interactive editing.

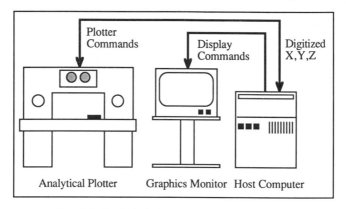

FIG. 2.3. Analytical photogrammetric workstation.

Gugan (1989) reviews the future development of photogrammetry. We will highlight a number of emerging trends here. Firstly, the dynamic optical superimposition of digitized image is gaining popularity. This allows the operator to view the result of digitizing over the stereo model in real time. Secondly, analytical plotters designed for different applications and offering a wider range of precisions have also been developed. Less expensive analytical photogrammetric workstations based on personal computers (Carson 1987) will further advance the use of photogrammetry for cartographic data capture.

Lastly, digital stereoplotters allow measurement from digitized aerial photographs and satellite images (Gruen 1989). Digital photogrammetry offers several advantages (Helava 1988), among them are that automated feature recognition can be performed more easily and digital images, unlike films which can become distorted, are stable.

Photogrammetric systems employing digital techniques are available (Cogan et al. 1988). Purely digital systems, however, rely on the availability of aerial photographs in digital form which is still an exception today but is expected to become more popular with the development of solid-state cameras (Gruen 1988). Complete automation of the photogrammetric process has captured our imagination. Its realization depends heavily on research in image matching (Guelch 1988) which in turn relies on the total automation of feature recognition.

Remote Sensors

Remote sensing is based on the age-old practice of obtaining information about objects of interest from a distance. The human eyes are still the major devices for this purpose. Technological advances have merely developed other means and methods to supplement human vision. An early remote sensing technique which is still in active use today is photography. Remote sensing satellites now in regular service employ sensors which perform basically the same task of image recording, although data are collected from a much higher altitude, and digital instead of analogue signals are produced. Landsat and SPOT, to be described later, are two well-known series of satellites carrying remote sensors.

Objects can be identified from a distance because of the special way they emit or reflect electromagnetic energy for which the sun is the major source. The characteristic spectral response, or spectral signature, of an object provides us with clues as to its identity. Colour is one such clue. The use of electromagnetic waves beyond the narrow visual portion of the spectrum supply us with many more clues. For this reason, data from remote sensors are of interest not only to surveyors and cartographers, but also to many practitioners in environmental sciences. For example, these data have been used to monitor the ozone layer, to detect and assess oil spills, and to help control pest and disease in forests.

Figure 2.4 compares the spectral sensitivities of conventional and satellite remote sensors. Solid lines indicate the part of the spectrum which can be sensed. Some electromagnetic waves, such as those from 5.5 to 7.5 μm (10^{-6} m), are absorbed substantially by the atmosphere. Those parts of the spectrum, for example between 1.5 and 1.75 μm, at which atmospheric transmission is high are called windows. Naturally, the sensitivities of the satellite sensors are designed to match the most useful windows.

Remote sensors are either passive or active. A passive sensor records the emitted or reflected electromagnetic energy which originated from the sun. An active sensor carries its own source of electromagnetic energy and can hence operate by day or night. Synthetic Aperture Radar (SAR), carried by Seasat, a companion of Landsat, is an active sensor which records much longer wavelengths than the passive ones (Wise 1989). The longer electromagnetic waves emitted from a SAR have a much higher capability to penetrate the atmosphere, suspended water vapours, and even shallow water. Weakness of naturally occurring emissions in this spectral region requires the active sensors to supply the energy.

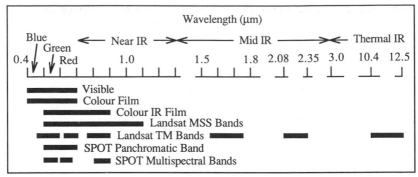

FIG. 2.4. A comparison of spectral sensitivities (modified from Lillesand and Kiefer 1987).

Both Landsat and SPOT carry passive sensors. Landsat-1, formerly known as ERTS-1 launched on July 23, 1972, was renamed in early 1975. This and a number of subsequent models are no longer functional, and Landsat-4 is gradually being phased out. Landsat-5, launched on March 1, 1984, is the only Landsat satellite fully operational today. It travels at a height of 705 km, sending data to several receiving stations around the world, including one in Prince Albert, Saskatchewan, Canada.

Another major resource satellite is the French designed SPOT (Systeme Pour l'Observation de la Terre) owned and operated by the Centre National d'Etudes Spatiales (CNES). Unlike the Landsat programme, the SPOT programme is a commercial venture. The first of its satellites, the SPOT-1, was launched on February 21, 1986. The SPOT-2, identical to SPOT-1, was launched in early 1990. Compared to the Landsat satellites, the SPOT satellites travel at a higher altitude of 832 km. There are two receiving stations for SPOT data in Canada: besides the Prince Albert station in Saskatchewan, the Gatineau station is outside Hull in Quebec.

There are four characteristics of remote sensors that concern cartographers. The first one is spectral resolution which describes the number and width of spectral bands recorded by the sensor. Sensors with a larger number of narrower bands offer better spectral resolution. An imagery with a higher spectral resolution allows better discrimination of objects. For example, black and white photographs providing a single band covering the visible portion of the spectrum are not as useful for discriminating objects than colour photographs providing a higher spectral resolution with three bands covering the narrower blue, green, and red portions respectively. Landsat-5 carries two types of sensors. The Multispectral Scanner (MSS) has four spectral bands. The other more advanced multispectral scanner, called the Thematic Mapper (TM), has seven bands covering a wider spectral range (Fig. 2.4). On board a SPOT satellite are two identical High-Resolution-Visible (HRV) sensors, each operating in either of two modes.

When used in panchromatic mode, one band generates black and white imageries. When used in multispectral mode, three bands are used (Fig. 2.4).

The second characteristic is spatial resolution which is the smallest separation of lines that can be detected. All other parameters being equal, sensors using shorter wavelengths offer better spatial resolution. Naturally, a sensor at lower altitude also produce better resolution. In addition, spatial resolution of civilian remote sensors are sometimes restricted by policies on national security. The MSS on Landsat-5 has a nominal resolution of 79 by 79 metres. The TM has a resolution of 30 by 30 metres except for the thermal infrared band which has a resolution of 120 by 120 m. For the SPOT, a resolution of 10 by 10 metres can be obtained in panchromatic mode, but a coarser resolution of 20 by 20 metres results in multispectral mode.

The third characteristic is radiometric resolution which describes the number of "grey levels" within the bands. Both the Landsat TM and SPOT provide 256 radiance levels while the Landsat MSS provides only 64.

The fourth characteristic is temporal resolution which is related to the time interval between successive coverage of the same area. A finer temporal resolution allows the detection of small changes in the object or phenomenon with time. Landsat-5 has a temporal resolution of 16 days. SPOT has a coarser resolution of 26 days. However, the sensors on the SPOT can be pointed away from the vertical to a maximum of 27 degrees on either side. This allows the temporal resolution to be reduced to 2.5 days for coverage at different angles.

Data from remote sensors are mainly used for change detections and thematic mapping (Ducher 1980). However, applications in topographic mapping at small scales are emerging. Konecny *et al.* (1987) showed that SPOT imageries promise good results for topographic mapping up to 1 : 25,000. Hartley (1988) described the use of SPOT-1 data for 1 : 100,000 mapping which realized significant cost reductions compared with conventional methods. Use of SPOT imageries for topographic mapping will increase the speed of producing the much needed global coverage of 1 : 50,000 topographic maps which in 1980 only cover about 40% of the earth's land area (Rhind and Clark 1988).

Besides the development of new satellites, such as the ERS-1 of the European Space Agency, the resolution of remote sensors will improve, particularly in the spectral and spatial aspects. Imaging spectrometers (Vane and Goetz 1988; Goetz *et al.* 1985) capable of recording 100 to 200 narrow bands are already being used on aircrafts. As for spatial resolution, it is conceivable that a higher resolution of 5 metres for civilian remote sensors can be achieved before the end of the century. Resolution even higher than that is possible in the future. For example, the civilian Hubble Space Telescope of NASA designed for astronomical observations has a theoretical ground resolution of 7 cm from a height of 275 km if modified to point towards the earth (Adam 1986). This resolution is enough to identify the type of vehicle.

Digitizers

A digitizer is different from other devices reviewed in that measurements are made on an existing map. They were the first major devices to capture cartographic data in digital form, and a lot of activities in computer cartography during the early years were centred around the conversion of existing maps to digital form. Early manual digitizing tables were electro-mechanical devices. The now popular solid-state versions have no moving parts and are extremely reliable. They have become standard equipment for almost all cartographic workstations. Other types of digitizers have been developed, including sonic ones which do not have pads, use wireless cursors, and are capable of three-dimensional digitizing (de Bruyne 1986).

The main developments in digitizing tables, sometimes called flatbed digitizers, have been in resolution, size, surface material, and sensitivity. Digitizers of very high resolution, for instance 0.01 mm (0.0004 inch), are available. However, for cartographic purposes, digitizers offering a resolution of 0.025 mm (0.001 inch) and an accuracy of 0.125 mm (0.005 inch) are sufficient because the tracking accuracy of a trained operator is about 0.2 mm. As for size, digitizing tables can now accept documents as large as 1.8 m by 7.3 m (72 inches by 288 inches). Changes in surface material results in some interesting products. Digitizing tables are available with translucent surfaces which can be used as a screen for the projection of an image from behind. The Numonics Gridmaster is a digitizing mat which is 0.8 mm (1/32 inch) thick and is flexible. Improvements in sensitivity now allows documents as thick as 1.3 cm to be used, and can also accept metalized electrostatic drafting paper or other conductive materials.

Manual digitizing is labour intensive. The tracking speed of a careful operator is about 1.5 mm per second. Using the 90 metres mentioned earlier as the length of convoluted lines on a low density topographic map, it will take at least 16 hours to digitize these lines. On top of this must be added the time to digitize text and the more regular lines such as roads.

Several approaches have been proposed to increase the speed and accuracy of line tracing. One is to use an intelligent cursor that can correct for tracking errors up to 1.6 mm (Hunka 1978). Another is to use an interactive (or semi-automated) line following system. There are several commercial products based on the interactive digitizing concept. One of them is from Laser-Scan Laboratories of Cambridge, U.K., which uses a laser and tilting mirrors to follow lines automatically. The optical tracking device locks on a line until it reaches a junction. Operator intervention is then required to restart the tracking. Another product is the Gerber Model

2500 video digitizing head which attaches to a Gerber pen plotter. This product was introduced in the early 1980s, and can be programmed to follow one of three instructions upon reaching a junction: turn left, turn right, or stop. An interactive digitizer offers some improvement in speed and is more accurate than manual digitizing. The main disadvantages are their cost and high degree of user intervention.

The truly automated digitizers today are raster scanners which are all based on the principle of breaking an image into pixels. Recent developments have produced many varieties of raster scanners, including kits to convert dot matrix printers and even pen plotters to scanners. A survey of low cost scanners for cartographic applications is given by Drummond and Bosma (1989).

There is also an increase in the number of specialized scanners, such as video digitizers (from Chorus Data Systems Inc.) producing raster images either directly from video tapes or through video cameras, digitizing cameras (from Datacopy Corp.) with a resolution of 3456 by 4472 offering 8 bits per pixel, and 35-mm slide and colour negative scanners (from Imapro) offering 16.8 million colours and a resolution of 4096 by 6000.

The resolution required for scanning cartographic data depends on the type of document. For reproduction of imageries such as aerial photographs, the normal screen used for halftone printing is 133 lines per inch (52 lines per cm). Investigation by Light (1986a) also shows that a resolution of 147 dots per inch (dpi) is adequate to depict the smallest identifiable objects (0.25 mm × 0.25 mm) on a photograph. Resolution requirements for line maps are different. The smallest colour-filled area on a map is about 0.5 × 0.5 mm, which seems to imply that a resolution coarser than 147 dpi is sufficient. In practice, at least 400 dpi is needed to depict the crisp sharp corners of areas and lines. There are raster scanners on the market that provide very high resolution of 2048 dpi.

Debate on the comparative merits of vector versus raster data has not been settled, and developments in raster-to-vector conversion techniques are still quite active (see chapter by Peuquet). Dedicated vectorizers are available but at a high cost. One of the main problems in raster-to-vector conversion is the automatic recognition of text, symbols, and line patterns. Text recognition systems are commercially available, some of which are sold as part of high volume data entry systems involving a raster scanner. However, text recognition in cartography is somewhat more complicated. Many different text fonts and sizes are often involved in a single map, and some of the text follows curves. Research such as that reported in Kahan et al. (1987) will help to solve some of these problems.

Storage Media

Digital data in either vector or raster format are voluminous, more so for raster data. We have mentioned earlier that about 1.8×10^6 coordinates are required to represent the lines in a low density topographic map. Assuming four bytes for each coordinate, a total of 7.2 megabytes are needed to store one map. Data reduction techniques using relative coordinates can probably reduce it to about 2 megabytes. The addition of other information such as text and road will increase the storage requirement to approximately 3 megabytes. Raster images cause a similar problem. A line map of 78.5×54 cm (31 inches by 21 inches) scanned with a good resolution of 400 dpi results in about 104 million pixels. Even for a monochromatic map, this requires at least 13 megabytes to store. A study by Light (1986a) shows that a total of 10^{13} to 10^{14} bits are required to store the 54,000 7.5-minute standard topographic quadrangle maps at 1:24,000 in raster format covering continental United States and Hawaii with an area of 7.84×10^{12} m^2. This provides a multi-layered coverage of both line and image data at a resolution of about 147 dpi.

We have seen a steady rise in magnetic disk capacities during recent years (White 1983). High capacity hard disks with 600 megabytes and more are available for large microcomputers and even personal computers. Since the number of magnetic drives that can be connected to a computer is limited, the storage capacity has a direct effect on the number of maps that can be stored online.

The most significant breakthrough in storage capacity is the introduction of optical storage devices. Optical disks are much higher in density, typically offering a total of 800 megabytes for the 5.25-inch versions, and 2 gigabytes for the 12-inch ones. A case for the use of optical disks for cartographic data is made by Light (1986b), and an application in mapping is described by Cooke (1987).

An optical disk uses a high-power laser beam to leave a permanent pit on its otherwise smooth surface. To read the disk, a low-power beam is directed towards the surface. The controller reads a zero if the beam hits a pit and is deflected, and reads a one if it hits a smooth surface and is reflected. Optical disks with prerecorded data that can only be read by the user are called CD-ROM (Compact Disks—Read Only Memory). Those that can be Written Once and Read Many times are called WORM disks.

CD-ROM and WORM disks cannot be erased. Although this nonalterability is sometimes regarded as an asset in data archival, it is a disadvantage in a dynamic environment. The use of magneto-optical technology produces Erasable Optical (EO) disks. Unlike normal optical disks, the laser beam does not leave a permanent mark on the surface of the EO disk. In writing mode, the high-power laser beam changes the polarity of dots on the surface with the help of an electromagnet. In reading mode, the effect of the electromagnet is temporarily shut-off, and a low-power laser beam senses the polarity of the dots. EO disks with capacities of 650 megabytes are available.

Progressively higher density drives for both magnetic and optical disks will continue to appear. Software techniques are also used to increase the capacity of disks. Drive maximizers, for example, integrate data compression/decompression algorithms into the disk drives. Depending on the file, a reduction of 20% to 800% is possible.

Conclusions

In this paper, we have reviewed several data capture devices and methods. Total stations and satellite positioning systems are used in the field. They can both generate positional data of high accuracy. Photogrammetry and remote sensing techniques leave permanent records of the earth obtained at high altitudes. The larger scale of aerial photographs makes photogrammetry the more accurate positioning method, but the holistic views provided by remote sensors offer a wider range of qualitative information. Digitizers transform hardcopy maps and imageries to digital form, and are essentially data converters. In this review, we have emphasized the technical characteristics of these techniques and their recent developments.

The most phenomenal advances in recent years have been in GPS and remote sensing, both space related. The rationale for using space technology for surveying and mapping is for the instrument to see (as in the case of remote sensors) and to be seen (as in the case of GPS satellites) easily. Future developments will improve the visibility, resolution, and accuracy of space sensors and transmitters.

Data capture and storage cannot be separated. Aerial photographs and satellite imageries have contributed to the explosion of databases in both analogue and digital forms. Advances in digital photogrammetry will substantially increase the proportion of digital data. High capacity, reliable, and cost effective storage media are therefore particularly important to cartography and GIS. We have reviewed new developments in storage technology, especially that for optical disks. Advances in hardware will increase the density of the storage media while advances in data compression and compaction techniques will optimize the use of existing space.

Acknowledgements

The author wishes to thank the following who have reviewed and offered valuable comments on previous drafts of this chapter: Dr. Eugene Derenyi, Professor Angus Hamilton, Dr. Alfred Kleusberg, Mr. James Secord and Dr. T. Y. Shih, all with the Department of Surveying Engineering, University of New Brunswick.

References and Selected Bibliography

Adam, J. A. (1986) "Verification: peace keeping by technical means, Part I: Counting the weapons", *IEEE Spectrum*, Vol. 23, No. 7, pp. 46–56.

Ashkenazi, V. and P. J. Summerfield (1989) "Rapid static and kinematic GPS surveying: with or without cycle slips", *Land and Minerals Surveying*, Vol. 7, No. 10, pp. 489–494.

Becker, R. (1988) "Very high resolution aerial films", *Photogrammetria*, Vol. 42, No. 5/6, pp. 283–302.

de Bruyne, P. (1986) "Compact large-area graphic digitizer for personal computers", *IEEE Computer Graphics and Applications*, Vol. 6, Mp/12, pp. 49–53.

Burgess, A. (1990) "The vulnerability and survivability of GPS", *GPS World*, Vol. 1, No. 1, pp. 46–48.

Carson, W. W. (1987) "Development of an inexpensive analytical plotter", *Photogrammetric Record*, Vol. 12, No. 69, pp. 303–306.

Cogan, L., D. Gugan, D. Hunter, S. Lutz and C. Peny (1988) "Kern DSP 1—digital stereo photogrammetric system", *International Archives of Photogrammetry and Remote Sensing*, Vol. 27, No. B2, pp. 71–83.

Cooke, D. F. (1987) "Map storage on CD-ROM", *Byte*, Vol. 12, No. 8, pp. 129–138.

Crossfield, J. K. (1989) "Evaluating efficient surveying technology for the land information system environment", *Surveying and Mapping*, Vol. 49, No. 1, pp. 21–24.

Dale, P. F. and J. D. McLaughlin (1988) *Land Information Management*. Oxford University Press.

Drummond, J. and M. Bosma (1989) "A review of low-cost scanners", *International Journal of Geographical Information Systems*, Vol. 3, No. 1, pp. 83–95.

Ducher, G. (1980) "Cartographic possibilities of the SPOT and Spacelab projects", *Photogrammetric Record*, Vol. 10, No. 56, pp. 167–180.

Energy, Mines, and Resources (EMR) (1976) "A guide to the accuracy of maps", Technical Report Series, Catalogue No. M52-46/1976, Surveys and Mapping Branch, Department of Energy, Mines and Resources, Ottawa.

Faig, W. and T. Y. Shih (1989) "Should one consider combining kinematic GPS with aerial photogrammetry?' *Photogrammetric Engineering and Remote Sensing*, Vol. 55 No. 2, pp. 1723–1725.

Faig, W., T. Y. Shih and G. Deng (1990) "The enlarger-digitizer approach: Accuracy and Reliability". *Photogrammetric Engineering and Remote Sensing*, Vol. 56, No. 2, pp. 243–246.

Ghosh, S. K. (1987) "Photo-scale, map-scale and contour intervals in topographic mapping". *Photogrammetria*, Vol. 42, No. 1/2, pp. 34–50.

Goetz, A. F. H., G. Vane, J. E. Solomon and B. N. Rock (1985) "Imaging spectrometry for earth remote sensing", *Science*, Vol. 228, pp. 1147–1153.

Green, G. B., P. D. Massatt and N. W. Rhodus (1989) "The GPS 21 Primary Satellite Constellation", *Navigation*, Vol. 36, No. 1, pp. 9–24.

Gruen, A. W. (1989) "Digital photogrammetry processing systems: current status and prospects", *Photogrammetric Engineering and Remote Sensing*, Vol. 55, No. 5, pp. 581–586.

Gugan, D. J. (1989) "Future trends in photogrammetry", *Photogrammetric Record*, Vol. 13, No. 73, pp. 79–84.

Guelch, E. (1988) "Results of test on image matching of ISPRS WG III/4", *International Archives of Photogrammetry and Remote Sensing*, Vol. 27, No. B3, pp. 254–271.

Hartley, W. S. (1988) "Topographic mapping with SPOT-1 data: a practical approach by the Ordnance Survey", *Photogrammetric Record*, Vol. 12, No. 72, pp. 833–846.

Hartl, Ph. (1989) "Remote sensing and satellite navigation: complementary tools of space technology", *Photogrammetric Record*, Vol. 13, No. 74, pp. 263–275.

Helava, U. V. (1957) "New principle for photogrammetric plotters", *Photogrammetria*, Vol. XIV, No. 2, 1957: 58, pp. 89–96.

Helava, U. V. (1963) "Analytical plotter", *Canadian Surveyor*, Vol. XVII, No. 2, pp. 131–148.

Helava, U. V. (1988) "On system concepts for digital automation", *Photogrammetria*, Vol. 43, No. 2, pp. 57–71.

Henderson, T. E. and C. A. Quirion (1988) "Use of GPS-derived coordinates in GIS environment", *Journal of Surveying Engineering*, Vol. 114, No. 4, pp. 202–208.

Hothem, L. D., C. C. Goad and B. W. Remondi (1984) "GPS satellite surveying—practical aspects", *Canadian Surveyor*, Vol. 38, No. 3, pp. 177–192.

Hunka, G. W. (1978) "Aided-track cursor for improved digitizing accuracy", *Photogrammetric Engineering and Remote Sensing*, Vol. 44, No. 8, pp. 1061–1066.

Kahan, S., T. Pavlidis and H. S. Baird (1987) "On the recognition of printed characters of any font and size", *IEEE Trans. on Pattern Analysis and Machine Intelligence*, Vol. 9, No. 2, pp. 274–288.

Kielland, P. and M. Casey (1990) "GPS cost benefit for hydrographic surveying", *GPS World*, Vol. 1, No. 1, pp. 40–45.

Kinlyside, D. (1988) "Some aspects on using GPS for airborne photogrammetric control", Vol. 49, December, pp. 55–72, *Australian Journal of Geodesy, Photogrammetry, and Surveying*.

Konecny, G., P. Lohmann, H. Engel and E. Kruck (1987) "Evaluation of SPOT imagery on analytical photogrammetric instruments", *Photogrammetric Engineering and Remote Sensing*, Vol. 53, No. 9, pp. 1223–1230.

Lee, Y. C. and G. Y. Zhang (1989) "Development of geographic information systems technology", *Journal of Surveying Engineering*, Vol. 115, No. 3, pp. 304–323.

Light, D. (1986a) "Mass storage estimates for the digital mapping era", *Photogrammetric Engineering and Remote Sensing*, Vol. 52, No. 3, pp. 419–425.

Light, D. (1986b) "Planning the optical disk technology with digital cartography", *Photogrammetric Engineering and Remote Sensing*, Vol. 52, No. 4, pp. 551–557.

Lillesand, T. M. and R. W. Kiefer (1987) *Remote Sensing and Image Interpretation*, 2nd Edition. Wiley, New York.

Logan, K. P. (1988) "A comparison: static and future kinematic GPS surveys", *Journal of Surveying Engineering*, Vol. 114, No. 4, pp. 195–201.

Maling, D. H. (1989) *Measurements from maps, principles and methods of cartometry*. Pergamon Press, Oxford.

McDonnell, P. W. (1987) "Data collector survey", *Point of Beginning*, August-September, pp. 28–38.

McDonnell, P. W. (1989) "P.O.B. 1989 total station survey", *Point of Beginning*, April-May, pp. 52–68.

Moore, H. D. (1989) "SPOT vs Landsat TM for the maintenance of topographical databases", *ISPRS Journal of Photogrammetry and Remote Sensing*, Vol. 44, pp. 72–84.

Mounsey, H. (Ed.) (1988) *Building Databases for Global Science*. Taylor and Francis, London.

Muller, J. C. (1987) "The concept of error in cartography", *Cartographica*, Vol. 24, No. 2, pp. 1–15.

Munjy, R., L. Fenske, R. Davies and M. Hartwig (1989) "Total station survey system (TSSS) software", *Surveying and Mapping*, Vol. 49, No. 4, pp. 173–178.

North Atlantic Treaty Organization (NATO) (1970) *Standardization Agreement, STANAG No. 2215*, Edition No. 3, Military Agency of Standardization.

Oimoen, D. C. (1987) "Evaluation of a tablet digitizer for analytical photogrammetry", *Photogrammetric Engineering and Remote Sensing*, Vol. 53, No. 6, pp. 601–603.

Paiva, J. V. R. (1990) "Critical aspects of modern surveying system", *Journal of Surveying Engineering*, Vol. 116, No. 1, pp. 47–56.

Remondi, B. W. (1985) "Performing centimeter accuracy relative surveys in seconds using GPS carrier phase", *Proc. 1st Int. Symp. on Precise Positioning*, Rockville, 15–19 April.

Rhind, D. and P. Clark (1988) "Cartographic data inputs to global databases", in Mounsey (ed.) (1988), pp. 79–104.

Reilly, J. P. (1989) "P.O.B. 1989 GPS equipment survey", *Point of Beginning*, June-July, pp. 24–38.

Slama, C. C. (Ed.) (1980) *Manual of Photogrammetry*, 4th edition. American Society of Photogrammetry, Falls Church, Virginia.

Tiller, R. (1990) "Surveying instruments today and tomorrow", *Land and Minerals Surveying*, Vol. 8, No. 1, pp. 29–33.

Vane, G. and A. F. H. Goetz (1988) "Terrestrial imaging spectroscopy", *Remote Sensing of Environment*, Vol. 24, No. 1, pp. 1–29.

Wells, D. (ed.) (1986) *Guide to GPS Positioning*. Canadian GPS Associates, Fredericton, Canada.

Wells, D., D. Delikaraoglou and P. Vaníček (1982) "Marine navigation with NAVSTAR/ global positioning system (GPS): today and in the future", *Canadian Surveyor*, Vol. 36, No. 1, pp. 9–28.

Wells, D. and A.. Kleusberg (1990) "GPS: a multipurpose system", *GPS World*, Vol. 1, No. 1, pp. 60–64.

White, R. M. (1983) "Magnetic disks: storage densities on the rise", *IEEE Spectrum*, Vol. 20, No. 8, pp. 32–38.

Wise, P. (1989) "Spaceborne radar imagery—its acquisition, processing and cartographic applications", *Cartography*, Vol. 18, No. 1, pp. 9–20.

CHAPTER 3

Developments in Equipment and Techniques: Microcomputer Graphics Environments

DAVID C. COLL

Department of Systems and Computer Engineering
Carleton University
Ottawa, Canada

Introduction

Cartography is a visual, graphical science in which immense amounts of geographically-based information are acquired, stored and processed, and are usually presented or displayed in some sort of graphical form (see chapter by Taylor). The nature and complexity of the data makes the application of the computer in cartography a very natural development. Computer systems and software packages have been developed over the last two decades to acquire, store, process, retrieve and display cartographic and geographic data; and to communicate them over digital telecommunications networks with ever increasing speed. In fact, computer cartography is becoming a well-established discipline in its own right. Immense amounts of relevant data from many different sources are available to the computer oriented cartographer. Hardware exists to store this data in a variety of forms, and software exists to create and manage the associated data bases, to process the data and to create cartographic products.

There are powerful special purpose computer systems whose software is dedicated to cartography and whose hardware is optimized to its support. The rate of progress in mainstream computer technology, however, is such that the performance of general purpose systems continues to provide more and more of the capability required for computer cartography. This is particularly true of the high resolution graphics capability associated with microcomputer based "personal" computers and engineering workstations. Use of general purpose computer workstations provides access to developments in operating systems and applications software as well as networking and communications. The latter is especially important as standards for broadband optical fibre local area networks are developed and the widespread introduction of Integrated Services Digital Networks becomes a reality.

Growth in computer cartography can be expected to accelerate as the power of microcomputer based graphics workstations continues to develop. The growth will be fuelled by improvements in the resolution, speed and colour rendering capability of displays, increases in the power of graphical and computational processors, the availability of large, high speed, low cost memories, expansions in the flexibility, range and level of operating systems and applications software, and reductions in the cost of powerful, microcomputer based graphics systems and workstations.

Thus, while computers are widely used for a variety of information processing purposes in cartography, the visual nature of most end-products makes it natural that computer graphics should be utilized for the creation and display of cartographic images.

Computer graphics, whereby an image is constructed under computer control and displayed on a cathode ray tube (CRT), provides a familiar and useful way to display computer output. Computer graphics are particularly useful for the presentation of complex data and in that context are universally used for the preparation and presentation of geographical and other information that is naturally graphical or pictorial in nature.

While computer graphics are as old as computers themselves, the emphasis in this chapter will be on microcomputer graphics, particularly on those associated with the IBM PC and compatibles.[1] The chapter provides a review of the principles behind microcomputer graphics in general and a description of current trends in personal computer and workstation graphics hardware and software. Following a brief historical review that introduces the various categories of computer graphics systems, their capabilities are defined in this chapter in terms of graphics controllers and processors, personal computer standards, add-in graphics and video cards, and graphics workstations and systems.

Microcomputer Graphics User Environments

The user of microcomputer graphics has a choice of working environments. As with all computer systems, a computer graphics display system has two major components: hardware and software.

---

[1] Other personal computers such as the Macintosh and Atari provide excellent graphics capability and superb user interfaces. The chapter in this volume by Raveneau, Miller, Brousseau and Dufour on the use of the Hypercard is an excellent example. However, PC-compatible graphics are emphasized because the open nature of the PC architecture has led to an abundance of third-party hardware and software products that provide the application developer with extensive possibilities not otherwise available on personal computers.

On the hardware side, the user has a choice of using the built-in microcomputer graphics, a separate plug-in graphics adapter that implements one or more of a variety of standard graphics modes, or a plug-in graphics system that implements the standards and a variety of graphics functions as well and provides a complete graphics display capability. All of the options depend on the presence of a display, a memory that determines the images and a processor that controls its contents. Any of the options may depend on conventional microprocessors for the generation of graphics and the control of the associated memory and display, or on special purpose graphics processors, the current versions of which are remarkably sophisticated.

On the software side, the user may program at the machine level with an assembler or compiler that incorporates graphics instructions, or with a library of assembler or compiler callable graphics functions and procedures, or may use applications software specifically designed for graphics or image processing.

Graphics System Components

Regardless of the particular environment selected, the graphics system normally works in conjunction with the central processing unit (CPU) and main memory of the microcomputer and consists of similar sub-components.

Hardware

As shown in Fig. 3.1, the graphics hardware has three components:

— the display, i.e. the CRT on which the image is displayed and the associated electronics which sweeps the electron beam(s) across the face of the tube. It is also known as a monitor, since television monitors are often used as the display device;
— the display memory in which the definition of the image is stored; and
— the display adapter which contains a processor that controls the display memory contents and the interfaces between the display memory and the display.

The display memory, also known as video memory, can be part of the directly addressable computer memory so that its contents may be modified directly by the CPU. In advanced graphics systems the architecture of the display memory is especially designed for use with raster displays and is usually part of the graphics system. The part of the display memory that determines the active display is called the frame buffer, so that what is actually displayed is a portion of, or a window into, a larger image.

The graphics adapter may, in fact, be the CPU of the microcomputer, or it

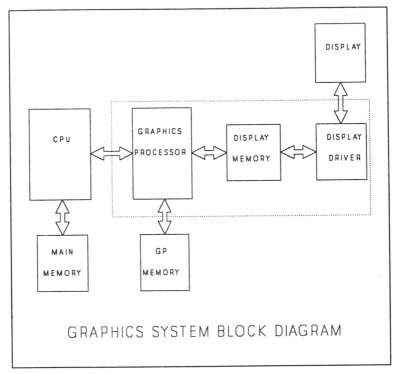

FIG. 3.1. Graphics system block diagram.

may be a simple alphanumeric controller that associates pixel values with character patterns and produces an appropriate video waveform to drive the monitor, or, in more advanced systems, it may be a full-blown graphics system processor and controller that "executes" picture drawing instructions and determines the display memory contents on its own. The graphics adapter may employ one or more general purpose microprocessors to accomplish its tasks or it may use graphics processor chip sets specifically designed for the task.

Software

In addition to the hardware, any graphics system incorporates software at a variety of levels. One level comprises the machine level graphics instructions of the graphics adapter itself. These control the display and the interfaces to the host computer and, when executed, create the image at the pixel level.

A second level of graphics software usually includes sub-programs that implement basic drawing functions or allow the programmer to exploit special graphics features of any particular system. High level languages usually include these device independent graphics functions and procedures as well as specific device drivers that allow the same high level software to be used with a variety of graphics systems. This software allows the development of applications level graphics software as a third level.

Historical Origins

Cathode ray tubes were used as output devices on the earliest computers due mainly to the fact that there were no other high speed output devices available, but also due to the fact that a graphic display, or plot, is a very useful way to present data to humans. Graphics displays were replaced very quickly as the normal output device on batch-operation-oriented mainframe computers by the line printer which produced "hard copy" output, numerical as it might have been. Printed output is essential in a batch-oriented environment because the user cannot interact with the output as it is being produced. On minicomputers, which were usually used interactively, the teleprinter (the printer part of the teletype machine) became the standard output device since the teletype machine was just about the only available digital input-output device. With the advent of the personal computer, however, the production of hard copy became less imperative and the CRT re-emerged in the form of a raster scanned monitor as the standard output device. In modern networked graphics workstation environments, very little hard copy output is produced and the graphics display is the major output device.

Modern computer graphics originated for technical and economic reasons[2] in two distinct but related ways: from the computer graphics processors that have been part of computers since their earliest implementations, and from the utilization of television-type CRT displays as the standard output devices on microcomputers and personal computers.

There are two basic ways in which the image is produced. The first is called **vector graphics**, and the second is called **raster graphics**. In vector graphics the position of the electron beam of the CRT is specified by the computer, and the beam is intensified or not depending on whether or not the locus of the beam is to be a visible part of the image. In raster graphics the beam is repetitively swept over the face of the CRT in a regular pattern from side to side and top to bottom, with the attributes of the image (brightness and colour) at each point determined by data generated by the computer.

---

[2] In the beginning there were no digital output devices other than teletypewriters that were cheap enough to be used with minicomputers. Line printers were developed, of course, but tended to be very expensive. The advent of TV-compatible character generators led to the introduction of the "dumb" video display terminals (VDT)—an idea that was picked up and used as the basic output device of personal computers.

Vector Graphics

CRTs were originally used in much the same way as an X-Y plotter is. That is, the position of the electron beam (the spot on the screen) was determined at all times by horizontal (*x*-axis) and vertical (*y*-axis) deflection voltages supplied to a cathode ray oscilloscope by the computer as numerical coordinates converted to voltages in analog to digital converters. Whether or not the CRT screen was illuminated at any point depended on the value of an intensity modulation (*z*-axis) voltage supplied by the computer. In this way, a point could be illuminated at any location of the CRT screen. A series of arbitrary points could be displayed by setting the coordinates of the first, letting the deflection voltages settle and momentarily intensifying the beam, setting the coordinates of the second point, letting the position stabilize, turning on the beam, turning it off and moving to the next point, and so on until all the points had been displayed. To see the display more than once the complete cycle had to be repeated, and to have a steady display the process had to be repeated without let up. The intensity of the display depended on the length of time the beam was intensified at each point. Unfortunately, as any one point was being displayed, all of the others faded away because of the nature of the phosphorescence of the CRT screen. Thus, every point had to be illuminated before it faded to such a level that flicker was apparent to the observer. Long persistence phosphors were used to reduce the flicker, but the resolution of the displays suffered greatly as a consequence and the rate at which the image displayed could be changed was reduced. Storage tube oscilloscopes were also used, but the image then became very static—but more on that later.

Simple point plotting was enhanced by the introduciton of line drawing hardware called vector generators. These are special purpose hardware which, when supplied with the coordinators of each end of a a line segment, could provide continuous voltages to move the electron beam in a straight line across the CRT screen. More restricted versions of the vector generator could provide drawing point movements of a set of fixed distances in a set of fixed unit directions (usually North, North East, East, South East, South, South West, West and North West) of various fixed lengths. As well, vector generators could draw a line from any point to any other point or from where the last segment ended to any point.

Thus, any image that could be drawn as straight line segments, i.e. as vectors, could be created on a CRT display. An image could be drawn by providing the vector processor with a list of vector end point coordinates and information to specify whether any segment was intensified or not. Gradually other features were added to the repertoire that the vector processor could perform, and the list of vector definitions became a list of display instructions that were executed by the vector graphics processor.

Vector graphics systems were highly developed as adjuncts to minicomputer systems by the early 1970s. In many cases graphics were generated by the graphics processor (GP) accessing a memory shared with the central processing unit (CPU) and executing graphics instructions placed there as the output of a CPU program.

For example, a program to draw a rotating wire frame rectangle might operate as follows. If the vertices of the rectangle were defined in a 4×2 matrix as:

$$
\begin{array}{ll}
X1 & Y1 \\
X2 & Y2 \\
X3 & Y3 \\
X4 & Y4
\end{array}
$$

the graphics program might consist of four instructions

DRAW_LINE_ABSOLUTE(X1,Y1,X2,Y2)
DRAW_LINE_RELATIVE(X3,Y3)
DRAW_LINE_RELATIVE(X4,Y4)
DRAW_LINE_RELATIVE(X1,Y1)

which would be sequentially executed continuously. The CPU program would consist of instructions to multiply the vertex matrix by a 2×2 rotational matrix, to modify the coordinate fields in the drawing instructions accordingly, and to increment the angle of rotation. The rate of rotation of the rectangle would depend on how often the CPU program executed, and the smoothness of the apparent motion would depend on the size of the increment to the angle of rotation.

The vector processor was essentially a "display-list" processor, moving the electron beam about the CRT screen according to a series of drawing commands (GP) generated by another program executing in the CPU.

Raster Scan Graphics

It was a small step to have the GP "write" the beam position to a memory with a one-to-one correspondence between the contents of the memory cells and the attributes of each location on the screen. The memory contents, when read in synchronism with a raster scanned CRT beam,[3] can be used to define the appearance of the display at each point.

---

[3] A raster scanned CRT is one in which the electron beam is swept horizontally from one side of the screen to the other, starting at the top of the screen; is returned to the starting side, but at a slightly lower position vertically, and is swept again to the other side. This horizontal sweeping is continued until the bottom of the screen is reached, at which time the entire process is continued from the top of the screen. When the image is changing rapidly, as it does in television, every other line is swept so that a complete image (with half the number of scan lines) may be created in half the time. The "missed" scan lines are then swept out on the next complete top-to-bottom cycle, so that the second set of scan lines is interleaved with the first. The complete raster is called a frame, and each of the interleaved partial rasters is called a field. The raster is re-drawn at a fixed rate. In standard North American television, there are approximately 30 frames, or 60 interleaved fields, per second. Each frame contains 525 horizontal lines. Thus, the horizontal sweep rate is about 15,750 lines per second.

The displayed image in raster computer graphics consists of a rectangular array of points or picture elements, called pixels or pels, whose intensity and colour are determined by data stored in a display memory. If the graphics system is operated in what is called graphics mode there is a one-to-one correspondence between each pixel and stored data that controls its appearance. When operated in text mode there is a correspondence between each pixel and preset character which determine its location depending on its relationship to the character of which it is part.

This chapter will concentrate on raster graphics, since virtually all modern microcomputer graphics systems use this system. In fact, modern graphics systems contain vector generators that create the data that determines the image on raster displays rather than drawing the image by controlling the beam position itself.

The resolution, and hence the quality, of a computer graphics image is determined by the number of horizontal lines in the raster and the number of pixels in each line. The number of pixels, together with the number of bits of information it takes to define each one, determines the size of the display memory and the speed at which it must operate. The image must be repetitively displayed at a rate that is sufficiently high to avoid flicker, which is very annoying to the observer. Thus, the rate at which the raster is created, i.e. the horizontal and vertical sweep rates, and the speed with which data must be retrieved from the display memory are determined by the resolution. For example, a raster with 480 visible lines repeated at a 60 Hz rate has a horizontal sweep rate of approximately 31 kHz. If there are, say, 640 pixels per line each with 16 colours each, the image occupies 153,600 bytes of display memory. When retrieved 60 times per second the data transfer rate is over 73 megabits per second, or a memory access time of about 54 nanoseconds per byte. Obviously, a random access memory of this size and speed would be very expensive. In fact, the secret behind the displays is that picture information is stored in specially designed video random access memories, called VRAM's, that have architectures explicitly designed to provide information in synchronism with a raster scan.

Most modern graphics systems are raster scanned, many with resolutions (and corresponding memories) that are sufficient to create very high quality images. Current microcomputer graphics systems have resolutions that vary from 320 by 200 two colour pixels to 1280 by 1024 26-colour pixels.

It was not always that way.

Raster Scan Alphanumeric Displays

When memory and wide bandwidth CRTs were very expensive, raster scan displays were rather crude. The earliest raster scan displays wre used for alphanumeric output on standard black and white television monitors; and the special-purpose integrated circuits, called graphics controllers, developed for this application were among the first large scale integrated circuits. Clever use of graphics controllers led to the use of graphics as the dominant output for personal computers. Consideration of how these early graphics controllers work provides insight into the principles behind all graphics systems.

Most personal computers have raster scan graphics that operate in either text mode or in graphics mode. In text mode, the screen is partitioned into character positions, for example into two dimensional arrays with 25 rows of 80 characters each. The actual characters displayed are drawn by the graphics controller or display adapter.

Several attributes of the characters displayed may be controlled. These include foreground (character) colour, background colour, brightness and blinking, and in monochromatic displays, underline and reverse video. In personal computers, every character on the screen is specified in display memory by a byte defining its value[4] and another defining its attributes. Thus, the 2000 characters displayed on a 25×800 text screen would be stored in a 4000 byte memory.

In the IBM PC, the memory available for use by the display adapters is part of the regularly addressable memory. The image itself may be directly manipulated by the CPU. This is called memory mapped graphics, and the ability of the central processor to modify images directly provides a very powerful graphics capability.

Every character that the graphics can draw is represented by a character box. The character box contains a two-dimensional array of binary digits (bits), say 14 rows each 9 columns wide. Each of the bits that is a logical 1 will correspond to a dot displayed in the foreground colour in the character. In this example, as shown in Fig. 3.2, the top two rows and the bottom row could be kept blank (logical 0) to allow for spacing between character rows, and the first and last columns could be kept blank to allow for spacing between characters. Characters would then be defined as a 7 by 11 array of dots, with the bottom two used for descenders on the lower case letters such as p, g or y, leaving the body of the characters drawn in a 7 by 9 array of dots. Different character fonts are created by different bit patterns within the character box.

Special drawing characters could occupy the entire 9×14 space so that adjacent characters could touch each other to create continuous lines and blocks.

Characters are drawn on a raster scanned CRT by turning the beam on and off according to the binary pattern representing the character box of each character in the character array. As the beam scans through the location on the screen where a particular character is to be drawn, the pattern for that character is retrieved and the beam is intensified or not depending on the

[4] With one byte used to define character values there can be as many as 256 distinct characters.

Column									
1	2	3	4	5	6	7	8	9	
0	0	0	0	0	0	0	0	0	Line Spacing
0	0	0	0	0	0	0	0	0	Line Spacing
0	X	X	X	X	X	X	X	0	Character Body
0	X	X	X	X	X	X	X	0	Character Body
0	X	X	X	X	X	X	X	0	Character Body
0	X	X	X	X	X	X	X	0	Character Body
0	X	X	X	X	X	X	X	0	Character Body
0	X	X	X	X	X	X	X	0	Character Body
0	X	X	X	X	X	X	X	0	Character Body
0	X	X	X	X	X	X	X	0	Character Body
0	X	X	X	X	X	X	X	0	Character Body
0	X	X	X	X	X	X	X	0	Descender
0	X	X	X	X	X	X	X	0	Descender
0	0	0	0	0	0	0	0	0	Line Spacing

FIG. 3.2. 9 × 14 character box.

logical value of the binary pattern representing that specific portion of the character.

The process is illustrated in Fig. 3.3. Three sets of addresses are maintained: the character position (the cursor position, actually) that points to the display memory which, in turn, points to the character box pattern position in a read only memory (ROM), and raster drawing point position that points to the bit position within the box.

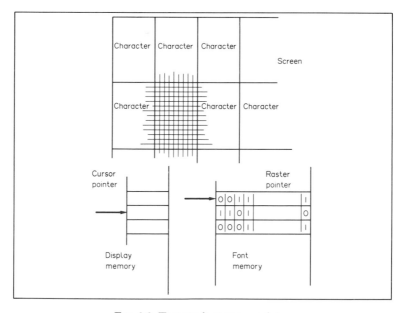

FIG. 3.3. Text mode memory pointers.

Personal Computer Graphics—the IBM Standards

Most personal computers utilize one of a set of *de facto* raster graphics standards with display attributes established by the IBM PC. The devices that drive the graphics are called display adapters in the PC world and the standards they have established are referred to as: the IBM Monochrome Adapter (MDA); the IBM Color Graphics Adapter (CGA); the Hercules Graphics Card (HGC), an extension devised to overcome severe deficiencies in the early IBM graphics; the IBM Enhanced Graphics Adapter (EGA); the IBM Video Graphics Array (VGA); and the IBM Multicolor Graphics Array (MCGA).[5]

These display adapters provide a range of resolutions and attributes and operate in a number of modes. These video modes, as they are called, are defined in Table 3.1.

The VGA has a resolution of 640 × 480 pixels with 16 colours. The MCGA is similar to the CGA but includes a 640 × 480 2-colour mode. Both have 320 × 200 256-colour graphics.

The text only Monochrome Display Adapter was meant to set the regular display standard. However, many users expressed a strong desire for graphics, which meant that they were saddled with very limited, low resolution CGA graphics. The need for higher resolution graphics was met by the Hercules card, an add-in card for the IBM PC and clones that provides 720 × 348 monochrome graphics as well as crisp text. Hercules graphics was accepted as a standard by software developers, and hence most popular software operates with the HGC as well as IBM-compatible standards.

The colour and brightness of CGA text can be defined. The attributes are coded in a byte as shown in Table 3.2.

The foreground, i.e. the character itself, may be any of eight colours, each of which may be normal or bright depending on the value of Bit 3. The background may be any of eight colours. The colours are coded as shown in Table 3.3.

The MDA text has attributes defined in Table 3.4.

The memory in the IBM PC is partitioned into 64K (1K = 1024 bytes) blocks. The first ten blocks, from 00000H to 9FFFFH, are assigned to ordinary user memory. The next, or A-block (A0000H to AFFFFH), and the next, or B-block (B0000H to BFFFFH), are devoted to the Extended Video Memory and the Standards Video Memory, respectively. Thus, 128K bytes are set aside for video memory.

The MDA requires 4000 bytes of memory starting at the beginning of the B-block. The pairs of bytes representing each character are stored

---

[5] VGA and MCGA were introduced with the IBM PS/2 series of personal computers and are implemented as built-in electronics in those machines. The others are implemented as add-in cards.

TABLE 3.1

CGA Video Modes

Mode	Type	Colour	Description
0	Text	No	40-column × 25 lines with no colour
1	Text	Yes, 16	40-column × 25 lines with colour
2	Text	No	80-column × 25 lines with no colour
3	Text	Yes, 16	80-column × 25 lines with colour
4	Graphics	Yes, 4	Medium resolution, 4 colour 320 pixels × 200 lines
5	Graphics	No	Medium resolution, no colour 320 pixels × 200 lines
6	Graphics	2	High resolution, black and white 640 pixels × 200 lines
7	Text	No	Monochrome text, 9 × 14 characters 80-column × 25 lines with attributes
8	Graphics	Yes, 16	Low resolution, 16 colour for PCjr 160 pixels × 200 lines
9	Graphics	Yes, 16	Medium resolution, 16 colour for PCjr, 320 pixels × 200 lines
10	Graphics	Yes, 4	High resolution, 4 colour, for PCjr 640 pixels × 200 lines
13	Graphics	Yes, 16	EGA Medium resolution, 16 colours 320 pixels × 200 lines
14	Graphics	Yes, 16	EGA High resolution, 16 colours 640 pixels × 200 lines
15	Graphics	No	EGA High resolution monochrome, for use with Monochromatic Display, 640 pixels × 250 lines
16	Graphics	Yes, 64	EGA High resolution, 64 colours for Enhanced Colour Display, 640 pixels × 350 lines
HGC	Graphics	No	High resolution monochrome graphics 720 pixels × 348 lines

TABLE 3.2

CGA Text Attribute Byte

Bit 7	Blinking	
Bit 6	Red	
Bit 5	Green	Background
Bit 4	Blue	
Bit 3	Intensity	
Bit 2	Red	
Bit 1	Green	Foreground
Bit 0	Blue	

sequentially in memory. Screen position is determined by the relative position of the byte-pairs in memory. In other words, the location of any data byte, relative to the start of display memory, may be computed according to the equation

TABLE 3.3

CGA Colour Attributes

Code	Red Bit	Green Bit	Blue Bit	Colour
0	0	0	0	Black (nothing)
1	0	0	1	Blue
2	0	1	0	Green
3	0	1	1	Cyan (green + blue)
4	1	0	0	Red
5	1	0	1	Magenta (red + blue)
6	1	1	0	Yellow (red + green)
7	1	1	1	White (red + green + blue)

TABLE 3.4

Monochrome Attributes

Bit 7	Blinking
Bit 6 + Bit 5 + Bit 4	Reverse
Bit 3	Bright
Bit 2 + Bit 1 + Bit 0	Normal
Bit 0	Underline

$$\text{offset} = 2 * ((80*\text{row}) + \text{column})$$

and the contents of the display memory at that location easily modified.

The CGA also uses 4000 bytes in text mode, starting at B8000. The CGA and the EGA store the data for more than one display screen at a time. The CGA memory can be partitioned into four independent display pages, any one of which can be selected for display. This allows the screen image to be changed very rapidly even though it make take considerable time to prepare any one of the four pages. The EGA may be programmed to have even more pages.

In graphics mode the amount of data stored per pixel varies with the graphics adapter. For example, the CGA requires 16K of display memory to define Mode 4 which provides 200 lines of 320 pixels each with 4 colours. There are 64,000 pixels, each requiring 2 bits to be defined, i.e. 128,000 bits or 16,000 bytes are required to define a screen in this mode. In Mode 6 there are 200 lines of 640 pixels each with one bit required to define the colour. In EGA mode 16, each of the 224,000 pixels requires 6 bits to specify its colour, resulting in a requirement for 1,344,000 bits or 168,000 bytes of display storage. This exceeds the 128K of available display memory, so that the EGA adapter has to perform extensive memory management.

In graphics mode, the data is stored in display memory in a manner compatible with the interleaved CRT raster. The data relevant to the first field of the raster is stored in continuous bytes in a continuous bank of memory, with the pixel data packed into the bytes. The data relevant to the second field is stored in a second continuous bank of memory locations beginning on an even 8K address boundary. Thus, memory locations are accessed sequentially when writing to the display, but they must be accessed non-sequentially when the image is being created. This causes a great deal of computational overhead.

In most colour CRT display drivers red (R), green (G) and blue (B) may be independently controlled, each with an eight-bit dynamic range. This implies that there could be 2^{24}, or 16,777,216 possible colours. Very few display memories are large enough to provide 24 bits to define the colour of every pixel, so that compromises have been devised. Having predetermined subsets of colours is one approach. These subsets are called palettes. For example, the four colours in Mode 4 are defined in terms of palettes. There are four predefined palettes with different combinations of four colours each. All include the background colour, i.e. the colour of the blank screen, which may be any one of the 16 possible colours defined in Table 3.5; the other three are fixed. For example, the CGA palettes defined in Turbo Pascal are listed in Table 3.6.

TABLE 3.5

Colours

Black	Dark grey
Blue	Light blue
Green	Light green
Cyan	Light cyan
Red	Light red
Magenta	Light magenta
Brown	Yellow
Light brown	White

TABLE 3.6

Turbo Pascal CGA Palettes

Palette	Colour 0	Colour 1	Colour 2	Colour 3
0	Background	Light green	Light red	Yellow
1	Background	Light cyan	Light magenta	White
2	Background	Green	Red	Brown
3	Background	Cyan	Magenta	Light grey

In high resolution, two-colour modes (e.g. Mode 6), the two available colours consist of the background colour and the drawing colour.

The pixel attributes that can be achieved with each of the various display adapters is related to the amount of storage available for each pixel in the display memory, but they also depend on the characteristics of the display monitor used with each. The MDA, CGA and EGA operate with digital monitors, while the VGA and MCGA operate with analog monitors. The digital monitors are driven by digital (on/off) signals while the analog monitors accept multi-level signals the same way a television monitor does. The Enhanced Colour Monitor that works with the EGA adapter has six binary (TTL) input signals representing primary and secondary red, green, blue. These six binary signals could define 64 colours, but the monitor is limited to 16 because only 4 bits are allocated in display memory for each pixel.

While a bewildering range of graphics adapters, display standards and associated monitors are used, the microcomputer user may be relieved of some confusion by the use of multi-standard display adapters, such as the ATI Technologies VIP Board. As well as displaying all VGA modes and attributes, this plug-in card runs EGA, CGA, MDA and Hercules software on analog, multisync, EGA and digital monochrome monitors. It also provides high resolution graphics (800 by 560 pixels) with a choice of 16 colours out of 256, and a variety of other modes that provide enhanced performance. The VIP adapter automatically detects the type of monitor connected and the mode being used. It is compatible with the IBM EGA so that previously written software will run with it installed.

Graphics Processors

To create an image in bit-mapped systems requires substantial computational effort. The attributes of every pixel that is to be different from the background colour must be calculated and the appropriate bits set in display memory.

For example, to draw a straight line from (x_1, y_1) to (x_2, y_2), the x coordinate may be stepped by unit amounts from x_1 to x_2 and the corresponding y coordinate calculated using the equation of a straight line:

$$y = m \times x + b.$$

The result is a real number, but the pixel coordinate must be an integer, so y must be truncated or rounded to the nearest integer. Other line drawing algorithms step x and y by fixed steps and calculate the closest pixel array point each time.

These computations are not only time consuming, but the resulting line is jagged because of the discrete nature of the display. Pixels may be thought of as discrete points in the display or as contiguous rectangular areas, with the pixel attributes determining the colour and intensity of that small area, which is sometimes called a tile. In this interpretation, a line drawing algorithm may treat the line as a figure with finite width and set the intensity of all the tiles that it crosses to be proportional to the amount of each tile covered by the line, thereby substantially reducing the jagged appearance of diagonal lines.

Creating pictures directly with the CPU keeps the cost of the graphics system low. However, the computational load on the CPU can result in very slow graphics and can cause severe interference with other tasks that the machine may be carrying out. As well, the CPU can be forced to wait while the contents of the display memory are being accessed for display, causing further delays. The introduction of associated graphics processors dedicated to graphics computation and driving the graphics display can alleviate many of these problems.

The GP is a processor that executes graphics commands or instructions stored in memory in exactly the same way that the CPU or host computer does. The display memory may be shared with the CPU or the GP may contain its own display memory.

The program that the GP executes is called a display file and comprises graphics instructions. The CPU can create or modify the display file and thus create graphic images. At one time graphics processors were extensive machines, equal in size to the host computers with which they were associated.[6] Today, although a GP may be as complex as the host computer, or even more so, it is usually implemented in VLSI. In personal computers, a GP is often contained on an add-in board.

Graphics processors are not new and systems such as the DEC VT-15 contained most of the features of modern GPs. The basic machine language instructions of the VT-15 consisted of character display: direct and from a string stored in memory; point and/or graph plot; display parameter controls, including light pen and display area edge interrupt handling; program sequence instructions (skip on conditions, flags, or sub-program name; jump and jump to subroutine); display status save/restore; basic vector drawing (eight possible directions, short or long); and arbitrary vector instruction (long and short). Line types and intensities, character and plot rotations, and blinking could be controlled. The display file was continuously executed to maintain a stable display on the associated monitor.[7]

The state of the art in graphics processors is now represented by the Texas Instruments TMS34010, its announced successor the TMS34020, and the Intel 82786 that are typical graphics processor chips. Their capabilities indicate the kind of operations utilized in graphics systems.

The TMS34010 is a microprocessor optimized for graphics system applications with data paths that are 32 bits wide, while pixels may be 1, 2, 4, 8 or 16 bits. It has general purpose programmability augmented with special

[6] For example the Digital Equipment Corporation Graphic-15 Display System, circa 1970, which contained the VT-15 Graphics Processor, occupied a 6-foot high cabinet as well as a large display console.
[7] The VT04 or VT07 were 17 inch and 21 inch CRTs, respectively.

graphics operations implemented in hardware that give it many advantages over hard-wired graphics controllers and general purpose microcomputers in speed and ease of use. Its programmability off-loads the host processor from the processing burden imposed by advanced drawing algorithms, graphics environment control and emulation of the plethora of graphics hardware standards. It works in conjunction with a number of other chips that go together to form a graphics processing system. These include a video systems controller, a colour palette and multi-port VRAMs. The 34010 provides separate interfaces to the graphics display, the host processor and its own internal memory. When the 34010 is operating as an associated, or slave, processor it transfers programs, data and graphics display lists between the host memory and its own local memory. Display and local memory refresh are performed automatically.

The instruction set of the TMS34010 includes graphics, move, general and program control instructions. All instructions and data used by the 34010 are stored in the local memory. The graphics instructions allow Boolean and arithmetic operations to be performed on corresponding pixels from two images or upon corresponding two-dimensional blocks of pixels. These pixel block transfer (PIXBLT) operations are an extension of bit block transfer (BITBLT) operations that apply to two level black and white images. The PIXBLT operations include all sixteen two-operand Boolean functions, add and subtract (with or without saturation), minimum and maximum. The 34010 executes floating point arithmetic operations by using special instructions and hardware. Line drawing, fill and move instructions are also included and drawing may be confined to preset windows.

Typical TMS34010 drawing speeds are listed in Table 3.7.

TABLE 3.7

TMS34010 Drawing Speeds

Horizontal line	25M bits/second
Average BLT move	25M bits/second
Line draw rate	1.25M pixels/second
IBM CGA text draw	43k chars/second

The TMS34020 is up to 50 times faster than the TMS34010. It has a closely coupled floating-point processor (the TMS34082) that is up to 10 times faster than other PC mathematics co-processors with a sustained rate of 40 million floating point operations per second (MFLOP). It incorporates 3-operand PIXBLT operations (source1, source2 and destination) at 142 megabits per second, 1.34 billion bits per second fill operations, and a 5 million pixel per second line drawing speed. The 34020 has more than 30 complex graphics instructions, allowing superior performance for 2-D and

3-D graphics applications.[8] While most mapping operations require planar graphics, the two- and three-dimensional functions available as hardware features on advanced graphics boards and workstations support motion in two dimensions, smooth shading, depth-cuing, true colour display and hidden-surface calculations for solids modelling applications important in digital 3-D terrain displays.

Graphics Support Software

The GP is accompanied by a set of software that consists of graphics routines callable from commonly used assemblers and compilers. This software produces code compatible with the particular GP and graphics display being used with it.

Software tools accompanying the TMS34020 include a C compiler, assembler/linker/debugger-simulator, mathematics, facsimile and windows management libraries, and a real time in-circuit emulator. This software runs on the IBM PC, the MAC II, on the DEC VAX, and on SUN and Apollo workstations.

With the advent of personal computers, most of which have a graphics capability, it is now commonplace for programming languages to contain extensive graphics commands. The compiler contains the high level language graphics statements as well as drivers for a wide range of graphics adapters. For example, all of the Borland languages (Pascal, C, Prolog, etc.) contain high level graphics statements which create code for the Borland Graphics Interface. The Interface, in turn, contains drivers for all common graphics adapters so the programmer need not be concerned with the niceties of the graphics drawing algorithms or the idiosyncracies of any particular graphics standard.

Typical of the graphics programming tools available to applications programmers are the graphics statements available with Borland Turbo C and Turbo Pascal.

Turbal Pascal incorporates text and graphics display procedures and functions ranging from direct access to video memory to polygon drawing, and a variety of adapter interface controls.

In this language the entire memory of the PC is defined as an array of bytes so that modifying the contents of the automatically displayed video memory area (which starts at B80000) changes what is seen on the screen. Text screens may be 25 or 80 columns wide and 25, 43 or 50 lines long, with colour

---

[8] These include arithmetic operations (absolute value, square root, add, subtract, multiply, dividend exponentiation); logical operations (AND, OR, XOR and compare); matrix operations (3 × 3 and 4 × 4 multiplies, transpose); graphics operations (polygon clipping, 2-D and 3-D linear interpolation and cubic spline); image processing (3 × 3 convolution); and vector operations (add, subtract, magnitude, scaling, dot and cross products, normalization and reflection).

characters or not. The TextMode procedure sets the mode. The cursor position may be set with the GotoXY procedure and read with the WhereX and WhereY procedures. The screen may be controlled with the ClrScr, ClrEol, DelLine and InsLine procedures. Windoes may be managed with the Window, OpenWindow, and CloseWindow procedures.

In graphics mode, Turbo Pascal (version 5) includes Borland Graphics Interface Adapter Drivers for CGA, EGA, VGA, Hercules, AT&T 400, IBM PC3270 and IBM 8514 graphics. Graphics control procedures include: InitGraph, which enters graphics mode; GraphDriver which identifies or sets the driver; GraphMode which sets the appropriate mode; and CloseGraph which resets the screen to the mode it was in when GraphInit was called. The current drawing point is found with the functions GetX and GetY. Individual points may be plotted with PutPixel and GetPixel. Lines are drawn with the procedures Line, MoveTo and LineTo. Lines may be drawn as solid, dotted, centre-dotted or dashed lines with normal or thick widths, determined by the SetLineStyle routine. Lines may be drawn "on top of" or merged with existing background. The graphics mode equivalent of the window is called a viewport and is established by SetViewPort. When Viewport is called with the parameter Clip enabled, output is allowed only to the area within the viewport. Viewports may be cleared with ClearViewPort. Colours are set by choosing a palette with GetDefaultPalette, or GetPalette; while palettes may be set with SetPalette or SetAllPalette or SetRGBPalette. Turbo Pascal allows shapes to be drawn with Rectangle, Bar (a solid rectangle), Bar3D, DrawPoly, FillPoly, Circle, Ellipse, Arc, PieSlice and Sector. Figures drawn with the FillPoly, Bar, Bar3D and PieSlice procedures are normally solid, however they may be filled with any one of 12 other patterns. Bounded regions may be filled with FloodFill. Text may be drawn in graphics mode during the OutText and OutTextXY procedures. Text is created from bit-mapped fonts stored as a 8×8 matrix of pixels, or from stroke-mapped fonts. Five fonts are available. They may be drawn horizontally or vertically with different sizes.

Complete Graphics Display Adaptors for the IBM PC

Typical of modern PC-compatible plug-in graphics adaptors using advanced graphics processor chips are the Matrox PG-1024 and PG-1281. These high-performance colour graphics adaptors convert IBM PC AT, RT or compatibles into high performance workstations. They provide 1024×768 or 1280×1024 display resolution with 8 bits per pixel on a flicker free 60 Hz non-interlaced raster. They are compatible with the VGA, EGA, CGA and PGA graphics standards and with the more general Virtual Device Interface and the Computer Graphics Interface. Associated accelerators for three-dimensional transformations extend their capabilities and speeds.

TABLE 3.8

Matrox PG-1024 Specifications

	PG-1024	*with SM-1024*
Display resolution	1024 × 768	
Frame buffer	1024 × 1024	
Displayable colours	16 or 256	256
Colour palette	4096	16,777,216
Grey levels	16 or 256	256
2D vectors/second	43,000	
3D vectors/second	10,000	80,000
Screen vectors	125,000	
Characters/second	30,000	
BITBLT/second	12,500,000	
Fill rate	25,000,000	
Shaded triangles/second		20,000
Gouraud rendering/s		10,000,000
Display list, Mbytes	1.5	2.5

TABLE 3.9

Workstation Specifications

SUN Workstations

	Sun-3/80	*SPARCstation 1*	*SPARCstation 300*
Processor	68030	SPARC	SPARC
Floating-point	68882	Built-in	Built-in
Main memory	4–16 Mbytes	8–16 Mbytes	8–40 Mbytes
Integer speed	3 MIPS	12.5 MIPS	16 MIPS
FP speed	160 kFLOPS	1.4 MFLOPS	2.6 MFLOPS
Mass storage	208 Mbytes	1.1 Gbytes	5.5 Gbytes
Monitors	16/19 inch		19 inch
Resolution	1152 × 900		1152 × 900
Colours	8/24 bits	8 bits	8/24 bits
2D vectors/second	325,000	400,000	450,000
3D vectors/second	100,000	175,000	200,000
3D polygons/second			5,500

DEC Workstations

	DECstation 3100	*VAXstation 3520*
Processor	RISC/FPU	CVAX/CFPA
Floating-point	Built-in	Built-in
Main memory	4–24 Mbytes	8–64 Mbytes
Integer speed	11–14 MIPS	5 MIPS
FP speed		
Mass storage	105 Mbytes	1.4 Gbytes
Monitors	15/19 inch	19 inch
Resolution	1024 × 864	1280 × 1024
Colours	1/8 bits	8/24 bits

The parameters of the PG-1024 listed in Table 3.8 are typical for this class of adaptors.

These systems are accompanied by an extensive library of graphics software and programming environment. The boards are supported by a number of CAD/CAM/CAE,[9] drawing and desktop publishing packages including AutoCAD, CADKEY, Computervision's MicroCADDS Geometric Construction and Detailing software, Graphic Software Systems CGI interface standard, GEM, various Tektronix emulators, MS-Windows and X Windows, Prior Data Services Intermaphics, Versacad, and Ventura Publisher, to name a few.

Workstations

Workstations are single user, desktop computers characterized by performance comparable to that of a super minicomputer.[10] They are relatively low in cost, with extensive graphics features and extensive software of all varieties. They are usually networked to a common file server and often work under a Unix operating system. Workstations are extensively used in research and development laboratories and engineering design environments. The first workstations were provided by SUN Microsystems and Apollo, and the major current providers are SUN, DEC, Hewlett-Packard (hp), IBM, Apollo, Intergraph, Tektronix and Silicon Graphics.

Until recently workstations were priced well above the microprocessor range but, as current technology provides personal computers with capabilities approaching that of low-end workstations, the decreasing price of workstations is approaching that of high-end personal computers.

Typical workstations are the SUN SPARCstation series and the Digital Equipment DECstation and VAXstation series, and some of their specifications are shown in Table 3.9.

Thus the basic machinery, hardware and software, is available (and in an advanced state of rapid development) to provide substantial improvements at low cost to current and future computer graphics applications.

Conclusion

In both cartography and GIS the pace of development in equipment and techniques will continue to have a major impact. There have been dramatic changes in the microcomputer graphics environment in recent years which have created exciting new possibilities for cartographers.

---

[9] Computer Aided Design, Computer Aided Manufacturing, Computer Aided Engineering.
[10] Performance is measured in MIPS (millions of instructions per second). The rule-of-thumb standard is the DEC VAX 780, which is a 1 MIPS machine.

CHAPTER 4

The Cartographic Workstation

BENGT RYSTEDT

National Land Survey of Sweden
Gävle, Sweden

Introduction

In this chapter it is intended to examine more closely the cartographic workstation (see chapter by Coll). The cartographic aspect of these will be stressed rather than their data acquisition and data analysis capabilities, and the assumption will be made that cartography is an integral part of a Geographic Information System.

The chapter will begin with a retrospective description of how computer assisted cartography started in the 1950s, and its early developments. After that the demands on a cartographic workstation will be presented. Finally, current technical equipment and a description of the key elements in the digital production line of National Land Survey of Sweden (NLS) will be dealt with.

History

I have often claimed that the first work in geographic information systems in its modern sense was done by Professor Torsten Hägerstrand at the University of Lund in 1955 (Hägerstrand 1955). He registered the coordinates for residential buildings on property maps at the scale 1 : 10,000 together with the real property designation, and combined that with data from the population register books kept by each parish. From this data he presented square grid maps by age groups and discussed how useful such data could be in urban and regional planning.

Hägerstrand's study inspired considerable research and development in Sweden (Nordbeck and Rystedt 1972), and in 1968 the Swedish Parliament decided that centroid coordinates would be registered for each real property and entered into the Swedish Land Data Bank System.

61

Today it is amazing to consider that the computer used in R&D on computer cartography for almost fifteen years was built at the University of Lund in the early 1950s. It was called SMIL (Siffermaskinen i Lund) and was equipped with a 4 K word primary memory, with each word of 40 bits' length forming a ten digit sedecimal number. In the program one word was divided into two half words where the first three digits were used for address numbers and the next two for instruction. SMIL had two registers, an accumulation register and a multiplication register.

An example of how it worked would be to add the two numbers stored in the cells with addresses 4AB and 4AC, with the following instructions: 4AB52 and 4AC50. The first instruction (4AB52) means that the contents of 4AB is moved to the accumulation register. The second (4AC50) means that the contents of 4AC will be added to the accumulation register. As this programming was very time consuming, it was a great relief when the first ALGOL compiler became available in the early 1960s. This meant that an ALGOL program was punched on to an eight channel paper tape and had to be read into the computer each time it was to be used. Input and output were handled with five channel paper tape which could be used to drive a typewriter and later a draughting machine.

A square grid map was rather easy to produce in the form of a matrix. A contour map, however, was more difficult. In the beginning we could only use the computer to calculate the values in the grid points, then transfer these to a sheet of paper to scale and draw the contour lines manually. Until cartographers had access to the use of penplotters for automated draughting, a technique to use line printers was widely used whereby by printing several letters on top of each other a grey scale was achieved. This technique became famous when it was used as part of the program package called SYMAP from the Harvard Laboratory of Computer Graphics. That was only twenty-five years ago, and since then has been tremendous development in the field which should continue with at least the same speed for another twenty-five years.

Demands on a Cartographic Workstation

Cartography is not a strong part of the computer graphics industry. It is only in the digitization part of map production that we are powerful enough to have any impact. In all other areas we must rely on developments in the computer graphics industry in general. We must follow technical developments in the computer graphics field in general, and develop cartographic software from that technology. That is why cartographic software has always lagged behind current technical developments. Nevertheless, it is important to establish the demands on a cartographic workstation.

By the term "cartographic workstation" I mean a set of hardware and software with the possibilities of input, editing and processing of cartographic data as well as final adjustment before plotting films for printing. If there is more than one workstation in the system they must all be connected as a network with a file server or host, where the geographic/ cartographic database is stored and maintained.

In general, the workstation should be built with standard components within the concept of open systems architecture. It should have a powerful processor for local processing, a high resolution graphics screen with at least sixteen simultaneous colours and the tools for entering data with a high degree of accuracy.

The normal tasks to be performed at a cartographic workstation would be to make high quality maps from different processed data residing in the host computer. This data may be acquired via manual digitization, scanning, the photogrammetic compilation or classification of satellite data. Hence there is a need for a versatile database management system and the capability to convert data from raster to vector structure and vice versa. The operator should be able to add data by digitization on the screen and on an attached digitizer.

All editing must be verifiable on the screen, and examples of editing functions are:

— parallel movements of lines;
— creation of point, line and area symbols;
— movement and rotation of symbols;
— text placement with the same size and form as in the final map;
— fit a curve to a series of points;
— select colours from a palette; and
— change grey scale (e.g. on orthophotos).

During the editing the result should be shown as WYSIWYG (what you see is what you get) and with possibilities to panorate in a whole map sheet. When the editing is finished there should be three options for output:

(1) that files be stored in the host computer;
(2) that plot files be available to a plotter for proof reading; and
(3) that colour separated plot files be available for final plotting on film.

The Technical Platform

When you are going to configure a cartographic workstation, you must be aware that it will be only one part of a production line. It must be built on the open systems concept, which will allow the inclusion of new technology as it becomes available.

Processing power is no longer a problem. There are several models on the market with enough power, and more are expected. Today a workstation with 10 MIPS (million instructions per second) is affordable and several vendors offer workstations in the neighbourhood of 25 MIPS. 100 MIPS workstations are anticipated in the near future. In the meantime, for many current users a PC with a 386 processor and a fast hard disk is satisfactory.

The graphic screen should have at least 1024×786 pixels and as a minimum 16 simultaneous colours out of a larger pallette. The trend is for graphic interfaces to do more and more work, faster and faster. It is easy to be amazed but difficult to choose the right graphic card, one that will support your software. It should have an image memory of at least 4 Mbyte, which is enough to store the content of a topographic map sheet in vector form. Such a card is hard to find today.

UNIX is preferred as the operating system since it provides networking and true multi-tasking. It helps to share data and devices such as plotters. Networking is of special importance since the cartographic workstation, sooner or later, will be integrated into the company's GIS. Facilitating the Network File System (NFS) should be a recommended *de facto* industry standard, and NFS software has already been adopted by some 270 vendors (Cosentino 1989).

The ideal cartographic software does not exist. CAD systems are, to a considerable extent, used for large scale maps and in general have no database management system. Small scale maps are complicated graphic products and the detailing and finesse desired by a cartographer are difficult to accomplish.

My conclusion is that development is proceeding so quickly it is dangerous to give specific recommendations. However, when a cartographic system is required, a request for proposals must be set up and perhaps be based on needs this chapter might inspire.

Cartographic Workstations at National and Land Survey of Sweden

NLS took its first steps on the road to computer assisted cartography in the late 1960s. The first items to be drawn by a draughting machine were grid nets and frames with coordinates. Development has continued and the mapping software has now been integrated into a system called AutoKa (automatisk kartframstallning—automated mapping). In 1987 NLS started development of a third generation of AutoKa with the following components:

— AutoKa-FC, a field computer to be used for field surveying and computing.
— AutoKa-PC, a PC-based workstation for large-scale mapping and geodetic measurements.
— Geodatabanken, a database management system with 3-D continuous geometry, topology and temporality.

In the second phase we are now achieving a digital production line by integrating the three systems mentioned above with Arc/Info and a high resolution laser plotter from Barco Graphics in Belgium. The main idea is that ordinary PCs can be used for the less demanding work such as manual digitization and vector editing. More demanding work such as name placement and final editing of small scale maps, powerful workstations should be used. When the production line is fully implemented, it is estimated that 500 PCs and 20 workstations will be needed in the organization.

With this distribution of computer power it is necessary to have strong management. There are rules on how to code a cartographic/geographic object, and how to store data in a Geodatabank where the database management system will take care of the administration. All small computers have to be attached to the host computer of NLS or to one of its five satellites. In this way we can secure accessibility to the geographic databases of NLS not only for our own use but also for external use.

Conclusion

The cartographic workstation, like many other aspects of cartography and geographic information processing, is dependent upon developments of both hardware and software in the computer industry in general, but considerable progress in the adaptation of existing equipment to the needs of cartography has been made.

References

Cosentino, Michael (1989) "Workstations versus PCs", *Professional Surveyor*, Vol. 9, No. 5, pp. 42–43.

Hägerstrand, Torsten (1955) "Statistiska primärupp-uppgifter, flyghantering och 'dataprocessing'-maskiner—ett kombineringsprojekt", *Svensk Geografisk Arsbok*, pp. 233–55.

Nordbeck, Stig and Rystedt, Bengt (1972) "Computer cartography. The mapping system NORMAP location models", *Studentlittertur Lund*, ISBN 91-44-04651-0.

CHAPTER 5

Methods for Structuring Digital Cartographic Data in a Personal Computer Environment[1]

DONNA J. PEUQUET

Department of Geography
The Pennsylvania State University
University Park, PA, USA

Introduction

Early digital cartographic storage formats were either hardware driven (e.g. a two-dimensional matrix that could be output directly on a line printer to produce a rudimentary graphic image) or a direct line-for-line transcription of the paper map. The need to understand the various options for cartographic data storage and the relative advantages and disadvantages of each one was superceded by the time, expense and overall difficulties of the initial conversion of map documents and other data into digital form. As a result, cartographic databases tended to be limited in size, regardless of the intended scope of the completed database or the amount of computer storage available. Increasing speed and capacity offered by new technology tended to outpace the growth in cartographic databases. Efficiency or flexibility problems of using an existing database were usually not encountered. When problems associated with storing or using the data were encountered, usually within the context of large, government-related systems, solutions tended to be ad-hoc.

This situation has changed dramatically in the past few years, with the advancement of data capture and input techniques, and the availability of large cartographic databases and remote-sensed imagery from government agencies worldwide. The rate of growth in database size has now far

[1]This chapter is based on an article published in 1984 *Cartographica*, Vol. 21, No. 4, pp. 66–113, entitled; "A conceptual framework and comparison of spatial data models".

surpassed growth in the hardware capacity to store and access the needed and available data on-line. Attempts to integrate new forms of data into existing systems have to-date proven extremely difficult, at best, and have spurred a significant amount of research on how to efficiently store and retrieve cartographic and other types of spatial data.

Nevertheless, computer hardware technology continues to advance at a very rapid rate (see chapter by Coll). A popular "small mainframe" machine introduced in the late 1970s was Digital Equipment Corporation's VAX 11/ 780, which could process about 1 million instructions per second and at that time could hold 2–4 million bytes of central memory. The cost of around $200,000, however, meant that although the needed computing power for handling real-world data sets and real-world applications were much more affordable than the larger, million-dollar-class mainframe, this still had to be used in a shared, multi-user environment.

Personal microcomputers became commonly available as packaged systems in the early 1980s, most notably with the introduction of the Apple computer. Nevertheless, this allowed many individuals access to computers for the first time. For many others, it meant that for the first time a machine could be dedicated to a single use. New problems were introduced simultaneously with new opportunities for utilizing computer technology. Because of the storage limits of these machines, their use in cartography and spatial data processing in general was limited to simple applications involving very small amounts of data.

It has only been in the last couple of years that these restrictions have been overcome. A new generation of desktop workstation machines, a sort of super-personal computer designed for scientific and office application as opposed to home use, have been introduced by a number of hardware manufacturers (see chapter by Coll). These workstation machines process 12 million instructions per second, allow up to 12–16 million bytes of central memory, can handle multiple high-density disk drives of over 700 million bytes, and are available for under $10,000. The speed and capacity of less expensive personal computers is also increasing very rapidly. In addition, the advent of CD ROM technology is beginning to provide a new method of distributing and storing large data files very inexpensively. This means that for the first time, dedicated personal computers also have the power and capacity to handle real-world data sets of significant size and complexity.

Advances in networking and telecommunications, as well as a dramatic increase in both total storage and total available computing power, have allowed sharing and trading of data files as never before to minimize duplication of effort. It is now possible for personal computers to talk to each other, to mainframes, or any other type of computer, worldwide. This has led to the expression: "The network is the computer." The existence of this capability (and potential for its abuse) has been dramatically publicized by the media when computer "hackers" have illegally accessed government, corporate and university mainframes. A distributed, shared environment is also at the expense of added problems of data conversion and understanding an ever-widening range of options employed by others.

All of these recent developments have had the effect of blurring the differences between mainframe and personal computing, with one very important difference: Selection, installation and management of all software and data files implemented on a personal computer is the responsibility of the individual user. Unlike the mainframe environment, there is usually no technical staff who can provide these services. More power, flexibility and choice comes at the expense of the need for more knowledge in order to obtain maximum benefit.

The remainder of this chapter attempts to address this need by presenting a general review of digital data models for the storage and manipulation of cartographic data applicable to the personal computing context. This discussion is organized in four sections. The first section provides an introduction to some general concepts and terminology that were developed within the field of Database Management Systems. The next section briefly introduces quality criteria that affect the decision of how data are to be digitally represented. This is followed by a discussion of the nature of cartographic data. The next section is the primary portion of the chapter and reviews the various types and relative merits of spatial data models that have been used in digital, cartographic data storage and processing applications, within a general taxonomic framework.

Basic Concepts and Definitions

A data model may be generally defined as an abstraction of the real world which incorporates only those properties thought to be relevant to the application or applications at hand, usually a human conceptualization of reality. A data model therefore tends to be tailored to a given application or problem context. Different users and different applications may have different data models. Each data model represents reality with a varying level of completeness.

More formal definitions of a data model have been given within the field of Database Management Systems. Within that literature, Ullman has defined the term "data model" as a group of entity sets together with the relationships among those sets. An entity is ". . . a thing that exists and is distinguishable; that is, we can tell one entity from another" (Ullman 1983). For example, a particular lake or a particular desk are each entities. An entity set is a class of entities that possesses certain common characteristics. For example, lakes, mountains and desks are each entity sets. Relationships include such things as "left of", "less than" or "parent of". Both entities and relationships can have attributes, or properties. These associate a specific value from a domain of values for that attribute with each entity in an entity set. For example, a lake may have attributes of size, elevation and suspended particulates, among others.

In both of the definitions above, the term data model is a very generic one, and indeed includes the paper map.

Within the field of Database Management Systems there is general agreement that to implement or evaluate a data model for use in a computer environment, the data need to be viewed at a number of levels in turn, progressing from reality, through the abstract, user-oriented information structure, to the concrete, machine-oriented storage structure (Klinger, Fu and Kuni 1977; Martin 1975). The purpose of deriving separate views is so that the progressive constraints of each view can be "factored-in" in a systematic and stepwise manner and also so that those constraints do not unduly bias the initial human conceptualization. This progression seems to logically fall into three clearly distinguishable levels.

As shown diagrammatically in Fig. 5.1, the term "data model" is used again here, but in a narrower context, limited strictly to denote the conceptual "information structure", without any concern for computer implementation. The term "data structure" is used to explicitly denote the human implementation-oriented view; i.e. a representation of the data model often expressed in terms of diagrams, lists and arrays designed to reflect the recording of the data in some computer language, such as Fortran or Basic. Finally, the "file structure" is the hardware implementation-oriented view that reflects the physical storage of the data on some specific computer storage medium, such as a magnetic disk.

Selection and Design of Cartographic Data Models: Form Versus Function

The process of implementing any specific data model design is therefore seen as a process of abstraction. Since no model or abstraction of reality can represent all aspects of reality, it is impossible to design a general-purpose data model that is equally useful in all situations. This is particularly true when dealing with complex phenomena. For example, some spatial data structures are good for cartographic display, but very inefficient for analytic purposes. Other data structures may be excellent for specific analytical processes, but may be extremely inefficient for producing graphics.

These are opposite extremes in the basic tradeoff involved in the data modelling process: The more perfectly a model represents reality (i.e. the more completely all entities and possible relations are incorporated), the more robust and flexible that model will be in application. The model will, however, be correspondingly larger and more complex. On the other hand, the more precisely the model fits a single application, excluding entities and

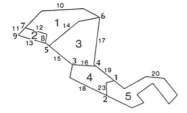

data model

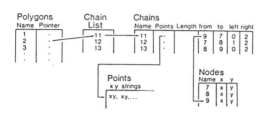

data structure

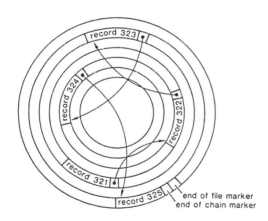

file structure

FIG. 5.1. Levels of data abstraction.

relations not required to deal with that application, the more efficient it will tend to be in storage space and ease of use.

The selection or design of a data model must, therefore, be based both on the nature of the phenomenon that the data represents and the specific manipulation processes which will be required to be performed on the data.

The process of deriving an optimum balance between these two positions is best accomplished in practice by utilizing both of these approaches simultaneously, balancing both toward what is hopefully an optimal compromise.

The performance vs. representational fidelity tradeoff just described impacts directly upon the storage, manipulative and retrieval characteristics of the data structure and physical file structure. It is necessary to examine these tradeoffs utilizing a specific set of usage-based criteria so that the overall quality or suitability of a specific data model can be evaluated within a particular context. The general criteria are:

(1) completeness
(2) robustness
(3) versatility
(4) efficiency
(5) ease of generation

Completeness may be thought of in terms of the proportion of all entities and relationships existing in reality which are represented in the model of a particular phenomenon. Robustness is the degree to which the data model can accommodate special circumstances or unusual instances, such as a polygon with a hole in it. Efficiency includes both compactness (storage efficiency) and speed of use (time efficiency). Ease of generation is the amount of effort needed to convert required by data in some other form into the form required by the data model.

In varying degree, each of these factors enter into consideration for any given application. The relative importance of each factor is, again, a function of the particular type of data to be used and the overall operational requirements of the system. For example, if the database to be generated is to be very large and to perform in an interactive context, compromises would likely be required with the first three factors because overall efficiency and ease of generation would predominate.

It is possible to quantitatively measure the performance of several of these criteria, such as speed and space efficiency for a particular data model. It is not possible, however, to provide quantitative measures for the more abstract factors of data completeness, robustness or versatility. Although quantitative efficiency specifications and empirical testing are beginning to be used to evaluate alternatives for large mainframe installations, experience and intuition will remain primary tools regardless of the size and complexity of the anticipated system.

Speed and space inefficiencies, even major and obvious ones, can often be tolerated if the total data volume is small or the data are infrequently used. In these cases, it is often less total effort to use extra computing time than to convert the data into a more efficient structure. Similarly, if the digital data can be obtained from elsewhere in a given format, the difficulty of initially generating the data in that format is not a concern.

The Nature of Cartographic Data

Geographic data are data which normally pertain to the earth. These may be two-dimensional, modelling the surface of the earth as a plane, or three-dimensional to describe subsurface or atmospheric phenomena. A fourth-dimension could also be added for time series data. When these data are organized in some form to create a model that can be visually interpreted, they can be called cartographic data. The current discussion will be limited to the case of two dimensions, since this is the basic case from which representations of higher dimensions are derived and is also by far the most common case.

There are several types of cartographic data, and the differences between them become obvious when they are displayed in map form, as shown in Fig. 5.2. The first is point data where each data element is associated with a single location in two- or three-dimensional space, such as the locations of cities of the United States. The second is line data. With this data type, the location is described by a string of spatial coordinates. These can represent either: (a) isolated lines where individual lines are not connected in any systematic manner, such as fault lines, (b) elements of tree structures, such as river systems, or (c) elements of network structures, as in the case of road systems.

The third type is polygon data, where the location of a data element is represented by a closed string of spatial coordinates. Polygon data are thus associated with areas over a defined space. These data can themselves be any one of three types: (a) isolated polygons, where the boundary of each polygon is not shared in any part by any other polygon, (b) adjacent polygons, where each polygon boundary segment is shared with at least one other polygon, and (c) nested polygons, where one or more polygons lie entirely within another polygon. An example of adjacent polygons are the state boundaries in a map of the United States. Contour lines on a topographic map are an example of nested polygon data.

A fourth category of data is some mixture of the above types. This might include different line structures mixed together, line structures mixed with a polygon structure or with discrete points. For example, in a map of the United States a state may be bounded by a river which is both a boundary between adjacent polygons as well as part of the tree structure of a river network. These four categories of spatial data are known as image or coordinate data (IGU 1975; IGU 1976). This means that these data portray the spatial locations and configurations of individual entities. A spatial data entity may be a point, line, polygon, or a combination of these.

Geographic data have a number of intrinsic characteristics which significantly differentiate them from other types of data. First, spatial entities have individual, unique definitions which reflect the entities' location in space. For geographic data, these definitions are commonly very complex, given the tendency of natural phenomena to occur in irregular, complex patterns. Particularly for geographic data, these definitions are recorded in terms of a coordinate system.

The relationships between spatial entities are generally very numerous, and, in fact, given the nature of reality or our perceptions of it, and the limitations of the modelling process, it is normally impossible to store all of them. The definitions of these relationships, and the entities themselves in

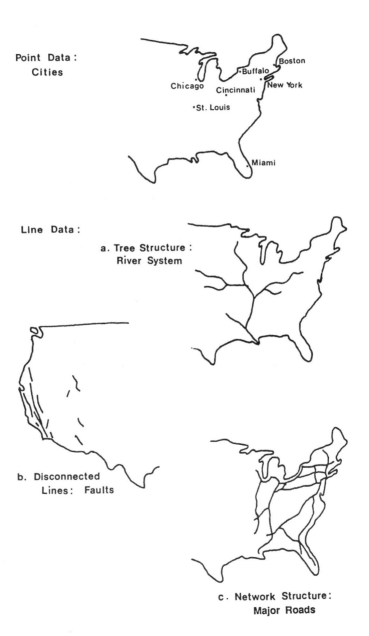

Point Data :
Cities

Boston
·Buffalo
Chicago   Cincinnati   New York
·St. Louis
·Miami

Line Data :

a. Tree Structure :
River System

b. Disconnected
Lines: Faults

c. Network Structure:
Major Roads

Polygon Data :

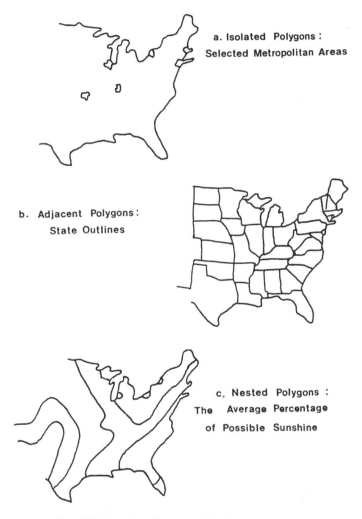

a. Isolated Polygons :
Selected Metropolitan Areas

b. Adjacent Polygons:
State Outlines

c. Nested Polygons :
The   Average Percentage
of Possible Sunshine

FIG. 5.2. Examples of cartographic data types.

the case of geographic data, also tend to be inexact and context dependent. This is true of even very basic spatial relationships such as "near" and "far", or "left" and "right".

An additional problem arises in the transformation of a conceptual data model into data structure and file structure views for computer implementation. Graphic input devices, such as digitizers, transform area, line, and point structures into numeric, computer-readable form by recording spatial coordinates of map entities. There is a basic problem underlying this transformation: Spatial data are by definition two- or three-dimensional, yet the coordinates must be structured in some way in linear or list fashion within the computer in some way that preserves the two-, three- or even four-dimensional relationships inherent within the data.

It is the combination of these intrinsic characteristics (multi-dimensionality, fuzzy entities and relationship definitions and complex spatial definitions) which makes the modelling of geographic data uniquely difficult. The models themselves tend to be complex and the resultant data files tend to be not very compact.

Basic Cartographic Model Types

A General Taxonomy

Geographic data have traditionally been presented for analysis by means of two-dimensional analog models known as maps (Board 1967). The map has also provided a convenient method of spatial data storage for later visual retrieval and subsequent manual updating, measuring or other processing.

Two additional classes of geographic data models have evolved for storing locational information in digital form; vector and tessellation models (cf. Fig. 5.3). In the vector class of data model, the basic logical unit in a geographical context corresponds to a line on a map such as a contour line, river, street, area boundary or a segment of one of these. A series of x-y point locations along the line are recorded as the components of a single data record. Points can be represented in a vector data organization as lines of zero length (i.e. one x-y location). Additional attribute, or descriptive, information about any given object can then be stored with the data record defining that object. With the polygonal mesh organization, on the other hand, the basic logical unit is a single cell or unit of space in the mesh. Additional attribute information concerning a given location can then be stored with the data record for that location. These two basic classes of data model are thus logical duals of each other.

Common usage has usually considered the two basic classes of spatial data models to be raster, or grid, and vector. As this discussion will show, however, the class of non-vector spatial data models encompasses much more than data models based on a rectangular or square mesh. This class includes any infinitely repeatable pattern of a regular polygon or poly-hedron. The term used in Geometry for this is a "regular tessellation". A tessellation in two dimensions is analogous to a mosaic, and in three dimensions to a honeycomb (Coxeter 1973).

Each of these approaches has also been used in fields other than cartography to represent spatial data, such as scanner images in picture processing. The characteristics of each of these approaches of models and

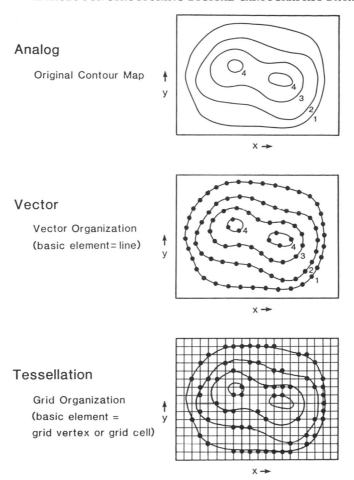

FIG. 5.3. Basic classes of geographic data models.

their tradeoffs for representing geographic phenomena should become clearer through the discussion of some specific examples that have become identified as the major types of data models within each of the two major classes of digital cartographic data models.

Vector Data Models

Most available vector data models can be easily categorized as one of the following "classic" types in terms of the mechanisms used for representing information. The first three types described below, the Spaghetti model, the Topologic model, and the Hierarchical Vector model represent a conceptual (and historical) progression of increasing information content and corresponding complexity. The fourth type, the Chaincode model, is actually a variant on the Spaghetti model but has special characteristics that warrant a separate discussion.

Spaghetti Model

The simplest vector data model for geographic data is a direct line-for-line translation of the paper map. As shown in Fig. 5.4, each entity on the map

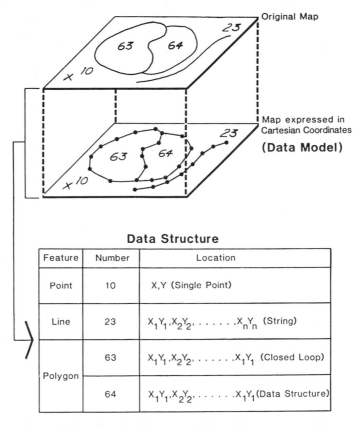

Data Structure

Feature	Number	Location
Point	10	X,Y (Single Point)
Line	23	$X_1Y_1, X_2Y_2, \ldots \ldots X_nY_n$ (String)
Polygon	63	$X_1Y_1, X_2Y_2, \ldots \ldots X_1Y_1$ (Closed Loop)
	64	$X_1Y_1, X_2Y_2, \ldots \ldots X_1Y_1$ (Data Structure)

FIG. 5.4. The "Spaghetti" data model (from Dangermond 1982).

becomes one logical record in the digital file, and is defined as strings of x-y coordinates. This structure is very simple and easy to understand since, in essence, the map remains the conceptual model and the x-y coordinate file is more precisely a data structure. The two-dimensional map model is translated into a list, or one-dimensional model. Although all entities are spatially defined, no spatial relationships are retained. Thus, a digital cartographic data file constructed in this manner is commonly referred to as a "spaghetti" file; i.e. a collection of coordinate strings heaped together with no inherent structure. A polygon recorded in this manner is represented by a closed string of x-y coordinates which define its boundary. For adjacent polygon data, this results in recording the x-y coordinates of shared boundary segments twice—once for each polygon.

The Spaghetti model is very inefficient for most types of spatial analyses, since any spatial relationships which are implicit in the original analog document must be derived through computation. Nevertheless, the lack of stored spatial relationships, which are extraneous to the drawing process, makes the Spaghetti model efficient for reproducing the original graphic image. The Spaghetti model is thus often used for applications that are limited to the simpler forms of computer-assisted cartographic display generation. Corrections and updates of the line data must rely on visual checks of graphic output.

Topologic Model

The most popular method of retaining spatial relationships among entities is to explicitly record adjacency information in what is known as a Topologic data model. A simplified example of this is shown in Fig. 5.5. Here, the basic logical entity is a straight line segment.

A line segment begins or ends at the intersection with another line or at a bend in the line. Each individual line segment is recorded with the coordinates of its two endpoints. In addition, the identifier, or name of each of the polygons on either side of the line are recorded. In this way, elementary spatial relationships are explicitly retained and can be used for analysis. In addition, this topological information allows the spatial definitions of points, lines, and polygon-type entities to be stored in a non-redundant manner. This is particularly advantageous for adjacent polygons. As the example in Fig. 5.5 shows, each line segment is recorded only once. The definitions and adjacency information for individual polygons are then defined by all individual line segments which comprise that polygon on the same side, either the right or the left.

The GBF/DIME (Geographic Base File/Dual Independent Map Encoding) model is by far the best known model which is built upon this topological concept. It was devised by the U.S. Census Bureau for digitally storing street maps to aid in the gathering and tabulation of Census data by providing geographically referenced address information in computerized form (U.S. Census 1969). Developed as an improvement of the Address Coding Guides, the initial GBF/DIME files were created in conjunction with the 1970 U.S. Census and used again for the 1980 Census.

In a GBF/DIME file, each street, river, railroad line, municipal boundary, etc., is represented as a series of straight line segments. A straight line segment ends where two lines intersect or at the point a line changes

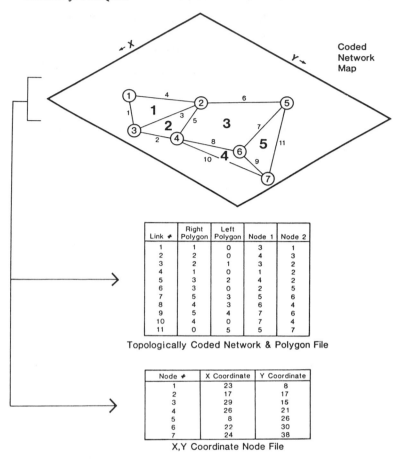

Fig. 5.5. The Topological data model (from Dangermond 1982).

direction. At these points and at line endpoints, nodes are identified (cf. Fig. 5.6).

As shown in Fig. 5.7, each GBF/DIME line segment record contains Census tract and block identifiers for the polygons on each side. The DIME model offers a significant enhancement to the basic topological model in that it explicitly assigns a direction to each straight line segment by recording a From-node (i.e. low node) and a To-node (i.e. high node). The result is a directed graph which can be used to automatically check for missing segments and other errors in the file, by following the line segments which comprise the boundary of each census block (i.e. polygon) named in the file. This walk around each polygon is done by matching the To-node identifier of the current line segment with the From-node identifier of another line segment via a search of the file. If line segment records cannot be found to

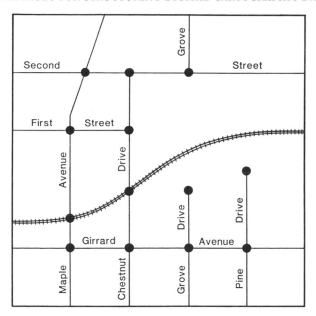

FIG. 5.6. Graphic elements of a GBF/DIME file.

completely chain around a polygon in this manner, a line segment is missing or a node identifier is incorrect.

Another feature worth noting is that each line segment is spatially defined, according to the definition of the model, using both street addresses and UTM coordinates. This is in recognition of the fact that some locational systems (e.g. street addresses), which may be needed for some types of applications, cannot be directly derived from conventional cartesian or polar coordinate systems.

A common problem with Spaghetti and Topologic models as described above is that individual records of entities do not occur in any particular order. To retrieve any particular line segment in a Topological model, a sequential, exhaustive search must be performed on the entire file. To retrieve all line segments which define the boundary of a polygon, an exhaustive search must be done as many times as there are line segments in the polygon boundary!

Hierarchical Vector Model

This type of model overcomes the very major retrieval inefficiencies seen in simpler topologic models by separately storing points, lines and polygons in a logically hierarchical fashion. Since polygons are comprised of the linear features that denote their boundaries, and linear features are comprised of strings of point locations, there are explicit links built into these models that

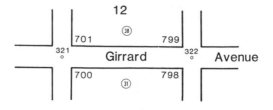

Street Name	Girrard
Street Type	Avenue
Left Addresses	701–799
Right Addresses	700–798
Left Block	38
Left Tract	12
Right Block	31
Right Tract	12
Low Node	321
X–Y Coordinate	155 000 – 232 000
High Node	322
X–Y Coordinate	156 000 – 234 000

FIG. 5.7. Contents of a sample DIME file record.

relate one type of feature to another, as shown in Fig. 5.8. These links also provide a direct retrieval mechanism. It is important to observe that Hierarchical Vector models normally (but not necessarily) also include topologic information.

One of the earliest example of a Hierarchical Vector model is POLYVRT (POLYgon conVERTer). POLYVRT was developed by Peucker and Chrisman (1975) and implemented at the Harvard Laboratory for Computer Graphics in the late 1970s. The general form of this specific model is given schematically in Fig. 5.8.

In POLYVRT, the term chain is used to denote the basic line entity. A chain is defined as a sequence of straight line segments which begins and ends at a node. A node is defined as the intersection point between two chains. The point coordinate information to define each chain is not stored as part of the chain record. Instead, a pointer to the beginning of this information within a separate Points file is recorded. Similarly, pointers are given within the Polygons file to the individual chains which comprise it. Note that the individual chain records contain the same explicit direction and topology information used within GBF/DIME; From- and To- nodes

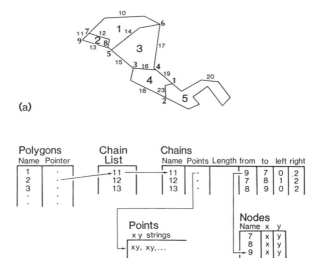

(a)

(b)

FIG. 5.8. (a) The POLYVRT data model. (b) The POLYVRT data structure.

as well as the left and right adjacent polygons. If a chain defines an outer boundary of the entire area, such as for chain 13 in Fig. 5.8, this outer area is denoted as polygon "0".

The Hierarchical Vector model provides a number of advantages over a Topologic model for retrieval and manipulation. First, the separation of the different classes of elements (i.e. polygons, lines, nodes and points) allows selective retrieval of specific classes at a time. For example, queries concerning the adjacency of polygons need only deal with the polygon and chain portion of the data. Only the individual chains which bound the polygons of interest are retrieved. The actual coordinate definitions are not retrieved until explicitly needed for such operations as plotting or distance calculations.

The number of line or chain records in a database using the Hierarchical Vector type of data model depends only upon the number of polygons present in the data and not on the detail of their boundaries. In computer implementation, this physical separation allows a much greater efficiency in needed central memory space as well as speed for many operations. This gives this type of model a significant advantage for use with entities which have highly convoluted boundaries. However, this physical separation also causes the need for a link or pointer structure. These non-data elements add a significant amount of extra bulk to the model. Pointers also represent a potential problem with assuring and maintaining data integrity. This is because incorrect pointers can be extremely difficult to detect or correct. The initial generation of this structure can also be cumbersome and time-consuming.

On the other hand, the Hierarchical Vector approach has considerable versatility. Peucker and Chrisman represent a POLYVRT data model and its corresponding data structure which are tailored to represent a set of adjacent polygons. The model can also be augmented for the representation of more complex data. This does not violate the basic concept of the model to add another level to the hierarchy, such as an additional level of polygons: Using this modified POLYVRT to represent a map of the United States, for example, the higher polygons could be states and the lower polygons could represent counties.

The TIGER (*T*opologically *I*ntegrated *G*eographic *E*ncoding and *R*eferencing) file, the data model to be used for the 1990 U.S. Census, is another example of a Hierarchical Vector model (Marx 1986). TIGER overcomes the lack of a retrieval mechanism of its predecessor, GBF/DIME, by storing separate directions for point, linear and polygonal structures, as well as other enhancements.

Chaincodes

Chaincode approaches are actually a method of coordinate compaction rather than a data model. They are included in this discussion for two reasons: First, this methodology provides significant enhancements in compaction and analytical capabilities and therefore has been frequently integrated into spatial data models, including some which will be discussed below. Second, Chaincodes have had a major impact on spatial data models and cartographic data processing to such an extent that it is commonly viewed as a data model in its own right.

The classical Chaincoding approach is known as Freeman-Hoffman chaincodes (Freeman 1974). This consists of assigning a unique directional code between 0-7 for each of eight unit-length vectors as shown in Fig. 5.9. The eight directions include the cardinal compass directions, plus the diagonals. Using this scheme to encode line data upon a grid of given unit resolution results in a very compact digital representation. As also seen in this example, *x-y* coordinate information need only be recorded for the beginning of each line. Direction is inherent in this scheme, providing an additional compaction advantage for portraying directed data, such as stream or road networks.

Through the use of special code sequences, special topological situations such as line intersections can be noted. One of the special coding sequences is also used for providing a mechanism for run-length encoding. This eliminates the need for repeated direction codes for long, straight lines. The flag used to signal that one of these special codes follows is "04". This directional chaincode sequence would mean that the line retraces itself, a meaningless sequence in most cases. It thus can be used as a convenient flag. The reader is referred to Freeman for his complete listing. These codes can, of course, be augmented or changed to suit a particular application.

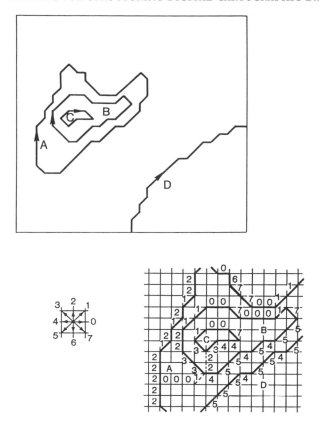

FIG. 5.9. Computer map and the resulting chaincoded lines (adapted from Freeman 1974).

There have been several variations of this coding scheme derived. The first, also described by Freeman (1979) is to utilize a 4, 16 or 32 vector notation on the same square lattice. The four-direction, encoding scheme allows representation of each code with 2 instead of 3 bits, and is sufficient in cases where the data tends to consist of long lines which are orthogonal to one another such as in some engineering applications. Sixteen or thirty two-direction coding allows for more accurate encoding of arbitrary-shaped curves. This smooths out the staircasing effect introduced by the directional approximations necessary when fewer directions are used for encoding (cf. Fig. 5.10). Similarly, there is a direct relationship between the number of directional vectors and the unit vector length for any given desired encoding accuracy for arbitrarily-shaped lines. In terms of compaction, this obviously presents a tradeoff between the number of direction-vector codes required to represent a given line and the number of bits required to represent each code.

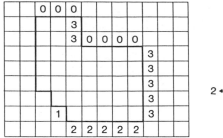

4 – directional coding

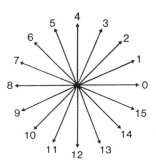

16 – directional coding

FIG. 5.10. Four- and sixteen-directional chaincoding schemes.

The primary disadvantage of chaincodes is that no spatial relationships are retained. It is, in fact, a compact Spaghetti model notation.

As previously mentioned, the primary advantages of the Chaincoding approach is its compactness. Chaincoding schemes are frequently incorporated into other schemes for the purpose of combining the compaction advantage of chaincodes with the advantages of another data model. The use of incremental, directional codes instead of cartesian coordinates results in better performance characteristics than the simple Spaghetti data model. The standard method of operation for vector plotters is to draw via sequences of short line segments utilizing (usually) eight possible direction vectors. Vector plotter hardware thus seems to be tailor-made for chaincoded data. Graphic output on these devices requires no coordinate translation, making the process very efficient.

Tessellation Models

As stated in the beginning of this section, Tessellation, or polygonal mesh models, represent the logical dual of the vector approach. Individual entities become the basic data units for which spatial information is explicitly recorded in vector models. With Tessellation models, on the other hand, the basic data unit is a unit of space for which entity information is explicitly recorded.

Grid and Other Regular Tessellations

All three possible types of regular tessellations have been used as the basis of spatial data models. Each has differing functional characteristics which are based on the differing geometries of the elemental polygon (Ahuja 1983). These three are square, triangular and hexagonal meshes (cf. Fig. 5.11).

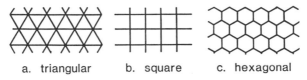

a. triangular   b. square   c. hexagonal

FIG. 5.11. The three regular tessellations.

Of these, the regular square mesh has historically been the most widely used primarily for three very practical reasons: (1) it is compatible with the array data structure built into the FORTRAN programming language, (2) it is compatible with a number of different types of hardware devices used for spatial data capture and output, and (3) it is compatible with cartesian coordinate systems.

In the earliest days of computer cartography, the only graphic output device commonly available was the line printer (Tobler 1959). Each character position on the line of print was viewed as a cell in a rectangular grid. Later devices for graphic input and output, particularly those designed for high speed, high volume operation, process data in rectangular mesh form. These include raster scanners, also known as mass digitizing devices, and color refresh CRTs. Remote sensing devices, such as the LANDSAT MSS capture data in gridded form as well (Peuquet and Boyle 1984).

The primary advantage of the regular hexagonal mesh is that all neighboring cells of a given cell are equidistant from that cell's centerpoint. Radical symmetry makes this model advantageous for analytical functions such as radial search and retrieval. This is unlike the square mesh where diagonal neighbors are not the same distance away as neighbors in the four cardinal directions from a central point.

A characteristic unique to all triangular tessellations, regular or irregular, is that the triangles do not all have the same orientation. This makes many procedures involving single-cell comparison operations which are simple to perform on the other two tessellations, much more complex. Nevertheless, this same characteristic gives triangular tessellations a unique advantage in representing terrain and other types of surface data. This is done by assigning a z-value to each vertex point in the regular triangular mesh (cf. Fig. 5.12). The triangular faces themselves can represent the same data via

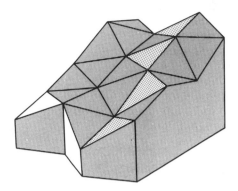

FIG. 5.12. A regular triangulated network representing surface data (adapted from Bengtsson and Nordbeck 1964).

the assignment of slope and direction values. Regular triangular meshes, however, are rarely used for representation of this type of data. Irregular triangular meshes are used instead, although Bengtsson and Nordbeck have shown that the interpolation of isorithms or contours is much easier and more consistent given a regular mesh (Bengtsson and Nordbeck 1964). Perhaps a contributing factor in the almost total lack of use of the regular triangular mesh for surface data is simply that such data are normally not captured in a regular spatial sampling pattern. An irregular triangular mesh has a number of other advantages which will be discussed later in this chapter.

Hierarchical Tessellation Models

Square and triangular meshes, as described above, can each be subdivided into smaller cells of the same shape, as shown in Fig. 5.13. The critical difference between square triangular and hexagonal tessellations in the plane is that only the square grid can be recursively subdivided with the areas of both the same shape and orientation. Triangles can be subdivided into other triangles, but the orientation problem remains. Hexagons cannot be subdivided into other hexagons, although the basic shape is approximated. These hexagonal "rosettes" have ragged edges (cf. Fig. 5.13). Ahuja describes these geometrical differences in detail (Ahuja 1983).

There are several very important advantages of a regular, recursive

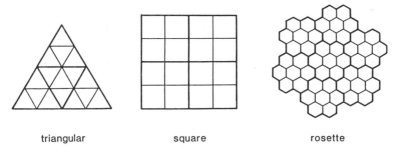

triangular   square   rosette

FIG. 5.13. The three regular tessellations in recursively subdivided form.

tessellation of the plane as a spatial data model. As a result, this particular type of data model is currently receiving a great deal of attention within the computer science community for a growing range of spatial data applications (Samet 1989a and 1989b). The most studied and utilized of these models is the Quadtree, based on the recursive decomposition of a grid (cf. Fig. 5.14).

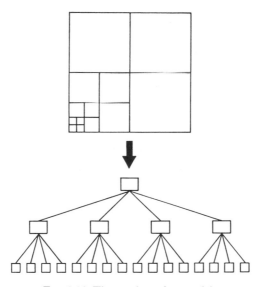

FIG. 5.14. The quadtree data model.

The advantages of a Quadtree model for geographical phenomena in addition to the advantages of a basic standard model include:

(1) recursive subdivision of space in this manner functionally results in a regular, balanced tree structure of degree 4. This is a hierarchical, or

tree, data model where each node has four sons. Tree storage and search techniques is one of the more thoroughly researched and better understood topics in computer science. Techniques are well documented for implementation of trees as a file structure, including compaction techniques and efficient addressing schemes.

(2) In cartographic terms, this is a variable scale scheme based on powers of 2 and is compatible with conventional cartesian coordinate systems. This means that scale changes between these built-in scales merely require retrieving stored data at a lower or higher level in the tree. Stored data at multiple scales also can be used to get around problems of automated map generalization. The obvious cost of these features, however, is increased storage volume.

(3) The recursive subdivision facilitates physically distributed storage, allows an easy mechanism for economically using central memory and greatly facilitates browsing operations. Windowing, if designed to coincide with areas represented by quadtree cells, is also very efficient. These are features which are very advantageous for handling a large database.

Advantages (1) and (3) also hold for the other two types of tessellations, taking into consideration that a recursive hexagonal tessellation has a branching factor of 7 instead of 4. Although all recursive tessellations can be viewed as having the variable scale property, the triangular and hexagonal versions do not have direct compatibility with cartesian coordinate systems. A comprehensive discussion of quadtrees and of its variant forms, as well as an extensive bibliography, has been given by Samet (1989a).

Irregular Tessellations

There are a number of cases in which an irregular tessellation holds some advantages. The three most commonly used types for geographical data applications are square, triangular and variable (i.e. Thiessen) polygon meshes. The basic advantage of an irregular mesh is that the need for redundant data is eliminated and the structure of the mesh itself can be tailored to the areal distribution of the data. This scheme is a variable resolution model in the sense that the size and density of the elemental polygons varies over space.

An irregular mesh can be adjusted to reflect the density of data occurrences within each area of space. Thus, each cell can be defined as containing the same number of occurrences. The result is that cells become larger where data are sparse, and small where data are dense.

Perhaps the irregular tessellation most frequently used as a spatial data model is the triangulated irregular network (TIN), or Delunay triangles (cf. Fig. 5.15a). TIN's are a standard method of representing terrain data for hill shading, landform analysis and hydrological applications as well as for

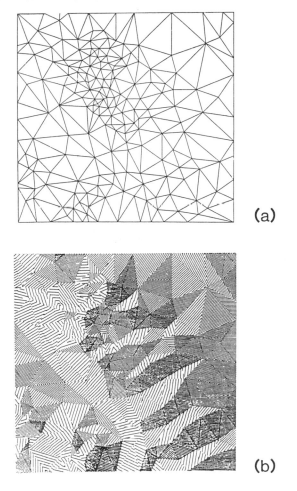

(a)

(b)

FIG. 5.15. A triangulated irregular network (TIN).

representing other continuous data, such as population density. There are three primary reasons for this: First, it avoids the "saddle point problem" which sometimes arises when drawing isopleths based on a square grid (Mark 1975). Second, it facilitates the calculation of slope and other terrain-specific parameters. Third, the data are normally recorded at points distributed irregularly in space.

A major problem associated with irregular triangulated networks is that there are many possible different triangulations which can be generated from the same point set. There are thus also many different triangulation algorithms. Any triangulation algorithm will also require significantly more time than subdivision of a regularly spaced point set. Another problem is that overlaying two irregular meshes where, for example each mesh represents a different data layer that are to be combined, is extremely difficult at best.

Scan-line Models

The Parallel Scan-line model, or Raster model, is a special case of the square mesh. The critical difference with the Parallel Scan-line model is that the cells are organized into single, contiguous, rows across the data surface, usually in the x direction, but do not necessarily have coherence in the other direction. This is often the result of some form of compaction, such as raster run-length encoding. This is a format commonly used by graphic scanning digitizers.

Although this model is more compact than the square grid, it has many limitations for processing. Algorithms which are linear or parallel in nature (i.e. input to a process to be performed on individual cells does not include results of the same process for neighboring cells) can be performed on data in scan-line form with no extra computational burden in contrast to gridded data. This is because null cells (i.e. cells containing no data) must also be processed in the uncompacted, gridded form. Many procedures used in image processing fall into this category. Other processes which do depend upon neighborhood effects, require that scan-line data be converted into grid form.

Vector versus Tessellation: Relative Merits

In summary, each of the two basic classes of digital data models have advantages and disadvantages which are inherent in the model itself. Individual models, such as the ones discussed above, can overcome these only to a limited degree, and always only by some sort of tradeoff. Vector data models are direct digital translations of the lines on a paper map. This means that the algorithms also tend to be direct translations of traditional manual methods. The repertoire of vector-mode algorithms is thus both well-developed and familiar within the cartographic realm. The primary drawback of vector-type data models is that spatial relationships must be either explicitly recorded or computed. Since there is an infinite number of potential spatial relationships, this means that the essential relationships needed for a particular application or range of applications must be anticipated.

Conversely, spatial interrelationships are "built-in" for regular, tessellation-type data models. Grid and raster data models are also compatible with modern high-speed graphic input and output devices. The primary drawback is that they tend to be not very compact. Regular tessellations tend to force the storage of redundant data values. Redundant data values, however, can be avoided by the use of a wide variety of compaction techniques such as run raster length encoding, analogous to chaincode run length encoding but used to record beginning and ending "runs" of successive locations along a single raster that have the same value. Another issue that used to be raised as a drawback was that the algorithm repertoire is less fully developed. This situation has improved significantly with experience in the use of Raster and other Tessellation-type models and adaptation of many techniques initially developed in the field of image processing for use with cartographic applications (Peuquet 1979).

From a theoretical perspective, Vector and Tessellation data models are logical duals of each other. The basic logical components of a Vector model is a spatial entity, which may be identifiable on the ground or created with the context of a particular application. These may thus include lakes, rivers, roads, and entities such as "the 20-foot contour level". The spatial organization of these objects is explicitly stored as attributes of these objects. Conversely, the basic logical component of a tessellation model is a location in space. The existence of a given object at that location is explicitly stored as a locational attribute.

From the above discussion, one can clearly see that neither type of data model is intrinsically a better representation of space. The representational and algorithmic advantages of each are data and application dependent, even though both theoretically have the capability to accommodate any type of data or procedure.

Complex Models

One approach to the intrinsic tradeoffs between Tessellation and Vector data models is to store the spatial data in (usually raster or grid) form, perhaps with only minor modification from its raw scanner raster output form. The data are then converted to vector form when advantageous for performing a given analytic or manipulative process. Frequently the result is then converted back again for graphic output. This conversion approach is the most commonly used because it is conceptually so straightforward. The problem with this approach, however, is that these data structure conversions can quickly become a bottleneck within a system (Peuquet 1981a; Peuquet 1981b). Tessellation-to-vector conversion requires some type of intricate line-following procedure, because cartographic lines are characteristically both convoluted and topologically complex. These conversion procedures represent significant systems overhead which must be avoided, or at least minimized.

A possible solution to this dilemma that has been suggested is the development and use of hybrid types of spatial data models which incorporate characteristics of both structures. The Vaster hybrid data model proposed by Peuquet (1983) is an example of such a model. This proved to be a potentially useful intermediate storage structure if both raster and vector-based processes were to be performed on the data, but few processes could be performed on it directly.

Other hybrid structures have also been proposed that are useful for specialized applications, but not as a basis for a flexible system that is to perform a variety of tasks. Perhaps the foremost example of this type of structure is the Strip Tree model that was originally proposed by Peucker (1976) and further discussed by Ballard (1981). The Strip Tree model is a method for representing map vectors by means of a hierarchy of bounding rectangles. This is a hybrid data model since the basic logical entity of the model is the cartographic line (i.e. vectors), but the lines themselves are not explicitly recorded. This representation allows such operations as line generalization, union, intersection and length of curves to be performed efficiently.

A more recent development in research on spatial data models has been the concept of Dual data models (Peuquet 1988). This is extremely simple in concept, and brings the raster vs. vector debate that began over ten years ago full circle. The basic idea of a dual model is that if vector-based and tessellation-based procedures are both needed in significant degree for a particular application, then *both* vector-based and tessellation-based data models should be used for storing the data. The major problem with this (and also a topic of current research), however, is how to use both within the same system in a balanced manner that simultaneously minimizes redundant data storage and format conversion. This is the basic difference between a dual model and maintaining two separate data models, as is currently done in some systems, which has the potential of storing information twice.

The discussion in this chapter was intended to present the primary options available for representing geographic data and to show how these options are conceptually interrelated. This chapter has dealt exclusively with data models that are explicitly designed for representing spatial (i.e. locational) information. The use of other types of data models for representing information pertaining to spatial phenomena, most notably the Relational Data Model that was developed within the field of Database Management Systems is unfortunately beyond what the limitations of space in a single chapter will allow.

References

Ahuja, N. (1983) "On approaches to polygonal decomposition for hierarchical image decomposition", *Computer Vision, Graphics & Image Processing*, Vol. 24, pp. 200–214.

Ballard, D. (1981) "Strip trees: a hierarchical representation for curves", *Communications of the ACM*, Vol. 24, pp. 310–321.

Bengtsson, B. and S. Nordbeck (1964) "Construction of isarithms and isarithmic maps by computers", *Nordisk Tidschrift for Informations-Behandling*, Vol. 4, pp. 87–105.

Board, C. (1967) "Maps as models", in *Models in Geography*, ed. P. Haggett, pp. 671–725, Methuen & Co., Ltd., London.

Coxeter, H. S. M. (1973) *Regular Polytopes*, Dover Publications, Inc., New York.

Dangermond, J. (1983) "A classification of software components commonly used in geographic information systems", *Proceedings*, U.S.-Australia Workshop on the Design and Implementation of Computer-Based Geographic Information Systems, pp. 70–91, Honolulu, Hawaii.

Freeman, H. (1974) "Computer processing of line-drawing images", *Computing Surveys*, Vol. 6, pp. 57–97.

Freeman, H. (1979) "Analysis and manipulation of line-drawing data", *Proceedings*, Nato Advanced Study Institute on Map Data Processing, Maratea, Italy, 1979.

I.G.U. Commission on Geographical Data Sensing and Processing (1975) Information Systems for Land Use Planning, report prepared for Argonne National Laboratory.

I.G.U. (1976) Technical Supporting Report D, U.S. Dept. of the Interior, Office of Land Use and Water Planning, Washington.

Klinger, A., K. S. Fu and T. L. Kuni (1977) *Data Structures, Computer Graphics, and Pattern Recognition*, Academic Press, New York.

Mark, D. (1975) "Computer analysis of topography: a comparison of terrain storage methods", *Geografiska Annaler*, Vol. 57a, pp. 179–188.

Martin, J. (1975) *Computer Data-Base Organization*, Prentice Hall, Inc., Englewood Cliffs.

Marx, R. W. (1986) "The TIGER system: automating the geographic structure of the United States census", *Government Publications Review*, Vol. 13, pp. 181–201.

Peucker, T. (1976) "A theory of the cartographic line", *International Yearbook for Cartography*, Vol. 16, pp. 134–143.

Peucker, T. and N. Chrisman (1975). "Cartographic data structure", *The American Cartographer*, Vol. 2, pp. 55–69.

Peuquet, D. (1979) "Raster processing: an alternative approach to automated cartographic data handling", *American Cartographer*, Vol. 6, pp. 129–139.

Peuquet, D. (1981a) "An examination of techniques for reformatting digital cartographic data, Part I: The raster-to-vector process", *Cartographica*, Vol. 18, pp. 34–48.

Peuquet, D. (1981b) "An examination of techniques for reformatting digital cartographic data, Part II: The vector-to-raster process", *Cartographica*, Vol. 18, pp. 21–33.

Peuquet, D. (1983) "A hybrid structure for the storage and manipulation of very large spatial data sets", *Computer Vision, Graphics and Image Processing*, Vol. 24, pp. 14–27.

Peuquet, D. (1988) "Representations of geographic space: toward a conceptual synthesis", *Annals of the Association of American Geographers*, Vol. 78, pp. 375–394.

Peuquet, D. and A. R. Boyle (1984) *Raster Scanning, Processing and Plotting of Cartographic Documents*, SPAD Systems, Ltd., Williamsville, N.Y.

Samet, H. (1989a) *The Design and Analysis of Spatial Data Structures*, Addison-Wesley, Inc., Reading, MA.

Samet, H. (1989b) *Applications of Spatial Data Structures*, Addison-Wesley, Reading, MA.

Tobler, W. (1959) "Automation and cartography", *Geographical Review*, Vol. 49, pp. 526–534.

Ullman, J. (1983) *Principles of Database Systems*, Computer Science Press, Rockville, Md.

United States Department of Commerce, Bureau of the Census (1970) "The DIME Geocoding System", in Report No. 4, Census Use Study, Washington.

CHAPTER 6

Mapping Software for Microcomputers

C. PETER KELLER

Department of Geography
University of Victoria
Victoria, British Columbia, Canada

NIGEL M. WATERS

Department of Geography
University of Calgary
Calgary, Alberta, Canada

Introduction

Over the last few years many cartographic software products have been developed for microcomputers. These products are diverse in capability, functionality, price and quality. New versions of available software continue to be released, and new programs continue to be developed. The objectives of this chapter are to introduce the reader to the microcomputer mapping software market, to provide an overview of software available for mapping on microcomputers, and to speculate briefly on market trends.

Given the diversity of available mapping products and the rapid growth of new software, it is infeasible to present a detailed review of the entire mapping software market. The intent of this chapter is merely to provide a general overview and not to rank the various products according to quality. Such value judgments can be obtained from published reviews, and ultimately are best left to the reader. Information summarized in this chapter was obtained from personal knowledge, from a survey of mapping software reviews published in a number of journals and from sales brochures.

The chapter commences with a brief overview of advances in micro-computing technology as related to mapping software. This is followed by a summary of the microcomputer mapping software market. Products available are differentiated by their operating system requirements and by functionality. Broadly categorized, it is possible to differentiate between software developed for IBM (and compatible) microcomputers, APPLE Macintosh microcomputers and "others". Functionality can be categorized broadly under painting, computer assisted drafting (CAD), thematic mapping, geographical information systems (GIS), image processing, contouring and "others". These categories and examples of relevant software packages are discussed in some detail.

The chapter concludes with a brief examination of possible future trends in the microcomputer mapping software market.

Microcomputing Technology and Mapping Software

The 1980s have experienced rapid advances in computing technology (see chapter by Coll), including the popularization of personal computing using microcomputers. The first microcomputers of the late 1970s and early 1980s had the reputation of being slow and cumbersome for computer mapping, with limited memory and poor display capabilities. Peripheral input devices, such as digitizing tablets, and output devices, such as plotters and printers, were extremely expensive and often lacked quality. Given the slow processing speeds resulting from the use of 8-bit central processing units and poor display capabilities, few, if any commercially worthwhile microcomputer mapping software emerged.

The digital industry has since advanced to produce microcomputers whose processing speeds and memory capabilities are beginning to rival the capabilities of old (and some contemporary) mini- and mainframe computers, making the distinction between these systems more and more difficult. To see just how far the field has progressed in a few years compare the pioneering work and images discussed in Walker (1985) with those available from today's technology.

Advances in microcomputing technology, in conjunction with declining prices, have popularized microcomputers during the 1980s to the point where they are becoming a regular feature in the workplace. Micro-computers have become an established teaching tool in most educational institutions, and they are an increasingly popular household item.

A driving force behind the popularization of the microcomputer has been the development of considerable volumes of software designed to satisfy industry, office, education and home demands. Software design efforts include the development of a diverse assortment of microcomputer mapping and related products, a development that was stimulated both by academic needs and the demands of the business community. Developments in microcomputer mapping software have taken various directions leading to products with very diverse capabilities, functionality, quality, price and hardware requirements.

Microcomputer mapping software has followed the trend of other microcomputer software developments with most software written to run on

IBM (and compatible) or APPLE Macintosh systems. IBM type products presently dominate in numbers. However, this is where any attempts at standardization stop (see also chapter by Evangelatos). There exists no agreed upon standard for coding spatial data, and most software packages require a unique data structure. The industry has not agreed on a single graphics standard, and different software products will require and support different hardware configurations and peripheral input, display and output devices. An introduction to some of the common file formats for microcomputer graphics packages is given in Waters (1989a). Although a single standard for microcomputer mapping is conceptually and operationally desirable such a goal may well be a chimera since the various formats which have been developed to date all have their unique strengths and weaknesses.

Most low budget mapping packages written for IBM and compatibles tend to operate on IBM XT type processors with standard RAM requirement. More expensive and sophisticated packages require IBM AT type architecture, or even a 386 processor. Available packages usually require 640 kB of RAM or less, although some of the latest releases of GIS and image processing systems specify additional RAM requirements. There is a tendency for more expensive systems to require the addition of a mathematical co-processor. Most IBM based packages require a color graphics monitor and a graphics board. Some software products require dual monitor workstations.

Originally hardware requirements for Macintosh software were simple. The Macintosh was a graphics oriented machine and no additional monitors and boards were necessary for processing although plotters and high quality printers were needed for hard copy output. Macintosh mapping software originally coalesced around the Mac Plus standard. This was limiting in that the machine was configured with a black and white monitor. Since that time, the Mac II series of modular machines (introduced in 1988 and since expanded to include the IIx, IIcx and IIci, see Waters 1989b), have made color available on the monitor with mapping software now taking full advantage of these new capabilities. Indeed some mapping software requires the extra memory and processing speed of the Macintosh II and will only run on this series of computers.

Microcomputer Mapping Software

The lists of software products discussed and summarized in this chapter was obtained from a number of sources. The authors' personal knowledge and collection of software information was complemented by a search for microcomputer mapping software reviews published in a number of journals. Notably the *American Cartographer*, the *Professional Geographer*, the *Journal of Geography* and *GEObyte* proved to contain useful mapping software reviews. Trade journals such as *PC Magazine*, *Macworld*, *MacUser* and the now defunct *MacGuide* also proved useful.

A database has been compiled of all the microcomputer mapping software identified for this review. The entries contained in this database are listed in alphabetic order in Table 6.1. This table by no means represents an all-inclusive list of available cartographic software and the list will not necessarily cite the latest version for any one product since these can change from month to month.

The database is stored in dBASE III Plus and Excel formats and will be made available through the software library of the Canadian Cartographic Association. It is intended to update this version of the list on a regular basis. The database contains the name of the product, the name of the company that markets the software, a classification of the software based on computer operating system and functional capabilities, price and where the software has been reviewed (if applicable). The reader is encouraged to obtain a copy of the database which can subsequently be customized, updated and expanded upon.

It was noted earlier that microcomputer mapping software is sold under many different names, depending on the type of cartography to be undertaken. It is possible to distinguish between painting software, object oriented drawing and computer assisted design (CAD) packages, thematic mapping programs, geographical information systems (GIS) software, contouring packages, image processing software, digitizing programs, educational packages, analytical software and a diverse assortment of other products.

The following sections will discuss the various categories of micro-computer mapping software introduced above, giving examples of products in the various price brackets. Unless specified otherwise, prices quoted are in U.S. dollars. The selection of products mentioned is somewhat arbitrary and reflects the knowledge of the authors and not necessarily the market leaders. The objective has been to draw upon a select number of products to highlight issues with which the reader should be concerned when considering the purchase of microcomputer based cartographic software. These sections will also discuss the products' strengths and weaknesses.

Painting

Painting packages support free hand sketching. The objective of the software is to let the user simulate various drawing media (for example, different size paint brushes) to paint images on the computer screen. Most painting programs also include template drawing capabilities. Painting programs generally assume that the screen imitates a single flat sheet of paper—the concept of painting on different layers for visual overlay is rarely supported.

The cartographer will find painting programs most useful in the design and layout stage of map production, when attempting to arrange and balance the visual components of the map, and when exploring figure-background relationships. Painting packages tend to give the user considerable choice of fill patterns, line weights and styles, and lettering fonts.

Paint programs were first developed for the Macintosh. Indeed when the original 128K "thin" Macintosh was introduced in 1984 it came bundled with just two programs: MacWrite, a wordprocessing program, and the then revolutionary MacPaint paint program with its strange assortment of tools such as erasers, paint brushes, paint cans and fill patterns which have since become industry generic concepts and standards. The venerable MacPaint program was quickly followed by a variety of third party programs which sought to provide enhancements in specific areas. These included Grey Paint, Full Paint, Super Paint and Pixel Paint among others. Some of these programs have become so popular and so essential for adding graphics to papers and articles that they are now bundled with wordprocessing programs. Super Paint, for example, the best selling paint program, is bundled with the latest version of Microsoft's wordprocessing program, Word 4.0. Paint programs on the Macintosh have also been used with "clip art" programs which provide map boundary files. One of the first of these was MacAtlas which originally provided boundary files for regional divisions of the United States. These files could then be transferred to a paint program by way of the Macintosh clipboard and then enhanced with cartographic details within the paint program. This provided one of the earliest forms of computer mapping on the Macintosh.

One of the more popular painting packages for the IBM and compatibles is PC-DRAW listed for $395. This package allows the user to draw on the screen in free hand or from templates. The software supports geometric functions like scaling, rotation and translation (see Myers 1985 for a complete discussion). Sechrist (1986) observes that the "speed at which drawings can be made rivals hand sketches, and they are of a quality rivaling drafted work". He concludes from his experience that the sales brochure's statement that "anything that can be drawn with paper and pencil can be drawn using PC-DRAW" is well founded.

An example of a more expensive painting option for IBM (and compatibles) is a package called VCN CONCORDE marketed for $695. This is advertised as an integrated drawing package which facilitates free hand screen painting with high quality text and animation including mapping functions. It has been reviewed in an article in *PC Magazine* published in 1986.

Computer Assisted Design

Computer assisted design (CAD) packages were developed predominantly for engineers, planners and architects to create digital plans. The strengths of these programs lie in template drawing and data entry. Spatial data is generally entered in absolute or relative cartesian or polar coordinate systems using a digitizing tablet or the keyboard. Attributes can be associated with points, lines and areas. Most CAD packages allow the user to draft on various layers which can be combined for visual overlay. Most also support some form of object oriented feature definition, that is a number of points, lines and areas can be combined to represent a single phenomenon. CAD software tends not to include map projections. Topology, that is the definition of spatial adjacency, proximity and hierarchy, generally is not given explicit consideration.

Cartographers will find CAD packages ideally suited for routine production of topographic maps and large scale plans. It is possible to use them as well for the design and production of thematic maps but, for design exploration, there are better and cheaper alternative software programs available.

CAD packages usually support coordinate geometry transformations, but rarely include cartographic query and cartographic analysis capabilities. The low cost packages usually lack flexibility in line weight, line style and lettering choice.

As has been noted, CAD packages were developed predominantly for engineers, planners and architects. The reader is advised to consult engineering and architectural journals as well as computing and software magazines for in-depth reviews of this type of software. The following is a brief introduction to some low cost computer assisted drafting options.

For IBM and compatible machines, at the low end of the budget, there are packages such as AUTOSKETCH listing for $99.95, EASYCAD listing for $169.95 and Generic Cad, versions of which may sell for as little as $49.95. Lipsey (1987) notes that drawing with AUTOSKETCH is simple and straightforward, but that "users with experience on more powerful machines and CAD programs will chafe at the bit". It would appear that a major drawback to AUTOSKETCH is the fact that drawings are kept entirely in memory while working on them. The amount of RAM available on the host platform will therefore limit the complexity of your drawing. McMaster (1989) notes that EASYCAD is an object oriented CAD system primarily for engineers and architects. He observes that its uses for mapping are questionable given that the software (Version 1.07) does not support line weight variations and only supports limited line styles.

Moving up in price a little will get you PRODESIGN II for around $300. This package offers a number of advantages over the previous two. Users can define their own symbol library and text fonts. Symbols and text can be placed on a map at any location, angle and size. Rudnicki (1987) notes that PRODESIGN II, although not a mapping package *per se*, opens new doors in microcomputer cartography at a most attractive price.

If money is not a consideration, then there are packages in the higher price bracket, such as FASTCAD and AUTOCAD (starting at well over $2000). These packages will yield considerable flexibility and additional capabilities, such as geometric transformations of objects and the ability to customize the software. Optional is the acquisition of additional macro language toolkits such as XP Tools and AUTOLISP for FASTCAD and AUTOCAD respectively.

CAD on the Macintosh has come a long way in a very short time. Peltz (1989) provides a recent review of this software noting that, at the time of writing, there were almost 40 CAD packages available for the Macintosh. Many of these packages are Macintosh versions of those available for the IBM-type microcomputers (or vice versa). Both Generic CAD and AUTOCAD are now available for the Macintosh. In these instances the genealogy of a program is often crucial. Programs conceived for IBM-type machines often fail to take full advantage of the Macintosh's graphical interface. This criticism was frequently levelled at the early Macintosh versions of Microsoft Word. Those ported in the other direction were often downgraded losing the attractive "look and feel" and ease of use on the IBM; Adobe's Illustrator is a case in point. Peltz (1989) provides a detailed review of three, low end, 2D CAD programs: Generic CAD, DREAMS and CLARIS CAD. The author states that historically, CAD programs were difficult to learn, difficult to use and costly. He goes on to note that all three problems largely have been resolved by most of the available Macintosh packages, and there would appear to be no excuse for not making an introduction to CAD a mandatory part of most cartography courses. In an aside to the main article, Peltz (1989) also discusses the high end, 3D CAD packages including the trend setting Microstation Mac from Intergraph. In an earlier review Peltz (1988) provides a more thorough account of the 3D packages, including a useful account of 3D CAD options. Peltz's review discusses image presentation procedures (wire frames, hidden line removal, shading, ray tracing and anti aliasing); projections (orthogonal and perspective); data input (coordinates, primitives, curved surfaces, multiview); 3D options (revolving, extruding, surface sweeping, connection of cross-sections, reshaping and joining) and image manipulation (rotation, translation and zooming). All the above expressions are likely to become standard terms in the lexicon of the computer assisted cartographer in the coming years.

The potential role of CAD packages for mapping on the microcomputer are summarized by Lipsey (1987). When reviewing AUTOSKETCH, he notes that this program was not designed as a mapping package, and that it would be unsuitable for many complex cartographic drawing tasks. This observation applies to most of the CAD packages. It can also be argued that CAD packages have many capabilities and options that are not required by a cartographer and that these systems are therefore in some ways awkward, over-sophisticated and too complex to be used for map design and production.

We will conclude by observing that, strictly from a drafting perspective, expensive CAD packages have the potential to yield high quality maps. The National Center for Geographic Information and Analysis (NCGIA) at Santa Barbara has gone as far as suggesting that a sophisticated CAD system, such as AUTOCAD, together with a state-of-the-art relational database system may be used as a rudimentary, vector based GIS. Low end, vector based GIS are few and far between and NCGIA has recommended that the AUTOCAD/dBASE combination can be used as an alternative to the higher priced vector based GIS, such as the ARC/INFO package. The MunMap GIS system is built around just such an arrangement.

Thematic Mapping

A number of microcomputer software products have emerged on the market that specialize in thematic mapping. Upon reviewing some of these products, one cannot help but get the impression that these programs were not designed for cartographers. Most packages do not require a sophisticated knowledge of computers or cartographic skills. Consequently, very few products offer the possibility of cartographic excellence. Most programs appear to have been designed as data exploration and display tools to be used by planners and consultants to assist in marketing, sales and site selection, tactical planning, forecasting and business profiling.

Few thematic mapping packages support digitizing capabilities. The user is generally required to purchase boundary files from the software vendor. However, most packages allow boundaries to be entered as x,y coordinate pairs using some form of text editor, an extremely awkward and slow method of spatial data entry. Some companies give the user the choice of purchasing additional software for digitizing.

These packages generally have severe limitations with respect to the design and placement of title, map body and legend. The capabilities to add comments, or to include readily a scale bar and an indicator of directionality are exceptions to the rule. Most packages, therefore, offer a map product adequate for data exploration, but not of sufficient quality for cartographic publication.

The bulk of all thematic microcomputer mapping packages concentrate on choropleth mapping. Dot density mapping is a popular second option while proportional symbol mapping rarely is supported. Isopleth mapping is only supported by a few vendors. It would appear that the computational complexities underlying the procedures of interpolation and smoothing have generally resulted in the marketing of independent software packages for isopleth mapping, commonly referred to as contour packages.

The most basic thematic mapping packages focus strictly on choropleth mapping. An example of a relatively inexpensive package for IBM (and compatibles) is POLYMAPS at $185. Lewis (1988) observes that this package, which comes with its own editor, contains very little cartographic flexibility with respect to design and lettering. He notes that documentation and error handling are poor. A step up in price will purchase something similar to QUICKMAP which retails for $295. This package also does not support digitizing but boundaries can be entered using a text editor. The program is completely menu driven and error handling appears to be comprehensive.

A popular choropleth mapping option is ATLAS*GRAPHICS at $449. This package allows users to produce a choropleth map in five minutes—providing that the boundary files and associated data are already in the computer. Documentation and error trapping are excellent, and the program has a comprehensive menu system. The package produces univariate and bivariate choropleth maps. With some skilful manipulation it is possible to produce aesthetically pleasing maps. A separate digitizing package (MAP EDIT or ATLAS*DRAW) is needed to digitize boundary files.

An example of a choropleth mapping package that supports digitizing is MULTIMAP at $495. Snaden (1986) comments that this package is not for the neophyte. He observes that DOS knowledge is a requirement. The package appears to allow for a reasonable amount of cartographic flexibility. Documentation is argued to be satisfactory.

A number of thematic mapping packages advertise themselves as programs designed for the business applications noted earlier. An example is RANDMAP, a choropleth and dot density mapping package listing for $495. Boundary files for the United States are company supplied. Another example is DIDS. This package falls into a higher price bracket at $1500 and still does not support digitizing.

MAPIT is a thematic mapping program that goes beyond simple choropleth mapping and can be obtained for the extremely competitive price of $95. Domier (1986) observes that this package can produce outline maps, conformant maps, interpolated maps and trend surfaces, reflecting to some extent the philosophy of the original SYMAP package. There appears to be considerable flexibility with respect to cartographic design. The program does not support digitizing, but the user can enter boundary files using a word processor. One of the greatest attractions of this package appears to be its price.

A more expensive thematic mapping package that goes beyond choropleth mapping is PC MAPICS retailing for £760 sterling. This package allows for the creation of choropleth maps, point maps, proportional symbol maps and line maps. It may be of interest to note that this program was produced by a company affiliated with the University of London, and the influence of cartographically literate authors appears to be reflected in the final product. Fisher (1988) observes that the package combines traditional business graphics (bar graphs, histograms, etc.) with mapping. The software has digitizing capabilities and there exists considerable cartographic freedom of design although the font options are limited.

Another more sophisticated thematic mapping package is DOCUMAP retailing for around $1000. This package produces choropleth maps as well as stepped surface maps. It contains interpolation routines, thiessen polygon divisions, and allows for the creation of digital terrain models. The software supports digitizing and some statistical capabilities. For example, it allows for the calculation of mean and standard deviation and JENK'S TAI statistic in the class break selection. The package has been criticized for being restrictive in design flexibility, especially with text limited to one size. DOCUMAP is available in a slightly modified version as MICROMAP.

Microcomputer software packages whose primary purpose is statistical analysis, but that support thematic mapping, include the SYSTAT/ SYSGRAPH combination, SAS/Graph and SPSS GRAPHICS. The last two packages were originally written for mainframe computing, with modules gradually becoming available on microcomputers. These programs allow the user to produce choropleth-, prism-, block- and surface maps. SAS/Graph and SPSS GRAPHICS require considerable effort and manipulation to achieve a pleasing map reflecting their mainframe lineage. Features such as map projections and coordinate transformation are supported. The main advantage of these two packages is their direct linkage to a powerful statistical package—their primary disadvantage is a high price.

There exist quite a number of other thematic mapping packages that are very specialized in their capabilities. For example the STATE DATA SYSTEM, selling for $20, is marketed as a data display package for the 50 states of America. Graphics are primitive and there is no flexibility in cartographic design. The program is available for both, Macintosh and IBM and has regression and scattergram capabilities.

One of the most interesting and academically satisfying choropleth mapping packages for the Macintosh is MacChoro. This package, developed by Image Mapping System of Omaha in 1986, takes full advantage of the Macintosh interface. The extensive series of drop down menus with numerous options allows the user complete control over the design of the final map. A large number of cosmetic features are available and the classification procedures include traditional methods such as quantiles, natural breaks, equal intervals and standard deviations as well as the newly popular unclassed method (see Mak 1989 for a discussion of the effectiveness of these competing methods of classifications). By the time of publication a new, color version of this popular package should be available.

Digitizing

A number of software packages have been developed to allow for the digitizing of boundary files. A good example of a digitizing package is ATLAS*DRAW. This package was developed to replace MAP EDIT, an early attempt at microcomputer digitizing software to create boundary files for the ATLAS*GRAPHICS thematic mapping package. Although developed as a complementary package to the ATLAS*GRAPHICS thematic mapping software, ATLAS*DRAW can be used to digitize outlines for other software packages using an import/export module. ATLAS*DRAW supports a number of map projections and various coordinate systems.

One of the lowest priced digitizing packages for microcomputers is ROOTS. ROOTS is a map digitizing program capable of building geographic databases for the ODYSSEY and MAP GIS software.

Scanners

An early review of Macintosh scanners from a geographer's point of view is provided by Dutton-Marion (1986). Dutton-Marion discusses several of the original models including Thunderscan, a flying spot scanner with a single photosensor, and the Micron Eye and Magic, each of which had an array of photosensors. Recent and comprehensive reviews of desktop scanners are provided by Beale and Cavuoto (1989), of color scanners by Bortman (1989) and grey scale scanners by Abernathy and Weiss (1989). The color scanners represent a remarkable technological development but Bortman (1989) suggests that color image formats and related software are still in the infancy of their development and those who can afford to wait should probably do so until these issues are resolved. The availability of this type of product represents an enormous opportunity for all those geographers concerned with the reproduction of images and maps. It is only reasonable to expect that this type of hardware will become as commonplace in geography departments and in the offices of professional geographers such as planners, as the mainframe digitizer was in the 1970s.

Contouring

As was noted earlier, few thematic mapping packages support spatial interpolation and isopleth mapping. One of the exceptions is DOCUMAP, a thematic mapping software product mentioned above that allows for the design of isopleth maps and the building of terrain models. The computational complexities underlying the procedures of interpolation, trend surface analysis, smoothing and terrain building have generally resulted in the marketing of independent software packages for management of the third dimension. These programs are commonly referred to as contouring packages.

Contouring programs can be differentiated by the interpolation techniques supported, by their ability to produce three-dimensional displays, and by the program's flexibility in allowing parameter manipulation. Some packages concentrate on triangulated terrain model definitions whereas others build altitude matrices. Sophisticated programs in the high price bracket tend to support profiling capabilities.

Examples of low priced contour packages for the IBM environment are CONTUR and SURFER, both retailing for $199. Goodchild (1985) notes that CONTUR is a contouring package that takes both, regular and irregular spaced data points, and that incorporates a triangulation algorithm for interpolation and a "bivariate" method to locate points between contours. The user has no explicit control over parameter definitions and inclusion of barriers is not supported. Labelling, line styles and axis definition are handled by a custom editor. Goodchild (1985) concludes that users familiar with the capabilities of more sophisticated spatial interpolation packages such as SURFACE II are likely to find CONTUR rather limited. However, the program is effective as a simple desk top system for generating quick contour maps from typed in data. Similar arguments hold for Golden's SURFER package although Version 2 of this program represents a major upgrade. Version 2 retails for $499 and allows for the creation of color contour and surface maps. The former may be superimposed on the latter giving the advantage of both types of presentation— location and visual impression, respectively. Contours can be smoothed and color zones can be created on both contour and surface maps. In addition, a wide variety of cosmetic features can be added, an option which was rarely available on the old mainframe packages such as SURFACE II.

There exist a number of other contour packages whose price generally reflects the number and complexity of algorithms supported, flexibility and user-friendliness. Some, such as the relatively expensive GWN-DTM package, include profiling capabilities.

At the time of writing there were relatively few contouring and surface fitting packages available for the Macintosh. One notable exception is MacGridzo. This program was originally ported from the MS-DOS world. It makes effective use of the Macintosh interface and uses the inverse distance method to produce an interpolated grid for contouring. Smoothing options are available for the contours.

A second package, GeoView, was released in 1988 and is a true Macintosh package, one that was not imported from the IBM environment. It has a variety of import/export options and data can be read in latitude-longitude formats or in UTM coordinates. It outputs contour, surface, overlay, trend surface, residual, volume and volume differential maps. It represents the high end of Macintosh contouring programs and retails for $900. Academic discounts are offered.

Macintosh users with non-existent budgets may be interested in two shareware products distributed by Somak Software Inc. which offers SurfaceGraph and Contour 81. The former makes three-dimensional surface diagrams while the latter, a more sophisticated program, generates both contour and three-dimensional surfaces on the Macintosh II series.

Geographical Information Systems

A more recent, but rapidly growing component of the microcomputer cartography market concerns microcomputer based geographical information systems (GIS). There appears to exist conceptual agreement with respect to what constitutes a geographical information system, but to this day we do not have an agreed upon written definition. GIS are therefore often confused with CAD packages, thematic mapping software and image analysis programs. The relationships between GIS and cartography are discussed in more detail by Taylor in Chapter 1 of this volume. To clarify the issue for our present purposes, it should suffice to say that a GIS is a program that allows for the entry, storage, display, manipulation, analysis and output of spatial data and its associated attributes. It should be able to handle point, linear and areal data as well as grid data structures. Ideally the spatial data should be stored in a topologically explicit fashion. A GIS may contain components of one, two or all of the computer assisted design, image analysis and thematic mapping programs.

GIS were originally developed on mini and mainframe computers, but advances in hardware technology now make microcomputers a viable operating option. Most microcomputer based systems have been written for IBM and compatible machines. Microcomputer based GIS are proving to be a lucrative enterprise, and a considerable number of mapping packages now are marketed as GIS. The consumer should be warned, however, that many vendors do not have a product that meets either the conceptual definition or the above written definition of a GIS. Microcomputer based GIS available today can be differentiated using the criteria of data structure philosophy, functional capabilities, processing speed, price and user-friendliness.

Cartographers will find GIS most useful when their work requires manipulation and analysis of the spatial database and associated attributes. Cartographers dealing with municipal type mapping will find vector based systems most appropriate, whereas those working predominantly with resource based data and remotely sensed images may prefer raster based systems. There appears to be a move by the software development industry towards the merging of the two data handling approaches within one system.

Microcomputer based GIS can be very expensive. A complete system including both hardware and software can start around $30,000, and it is not unknown for a user to spend over $100,000 on a fully equipped microcomputer based platform. There are a number of exceptions to this trend. Some universities have developed their own systems. Perhaps the best example is the IDRISI package developed by Ron Eastman at Clark

University (Eastman 1988). This package has found widespread acceptance and is continually being improved. It is a raster based GIS retailing for under $100 (academic price). Its flexibility, user friendliness and academic orientation make it excellent value for the money. Another exception is a product called MAP developed at the Harvard Laboratory for Computer Graphics and Spatial Analysis. The MAP system also is a raster based GIS which has been modified a number of times. It is now marketed under various names by several agencies including the Ohio State University which produces OSU MAP. Earlier versions include pMAP and aMAP.

One of the earliest attempts at developing software that would allow for spatial data entry, display and analysis was a package called UDMS (Urban Data Management System). This package, originally developed by Vince Robinson for the Z80 chip and the CP/M operating system, is available as public domain software and now runs on IBM and compatibles. Analytical capabilities are limited and concentrate on transportation analysis and the gravity model. Topological overlay is not supported. The package lacks graphics sophistication, but represents an excellent introduction to the topic of spatial analysis.

Examples of commercial geographical information systems in the $30,000 price range include ARC/INFO, PAMAP GIS, SPANS and TERRA-SOFT. These systems can be purchased in modules, each module costing a few thousand dollars. Most systems have a basic data entry module, a data manipulation and analysis module, an output module, an image analysis linkage module, a data exchange module and a terrain building module. Systems generally contain two major database managers, one for the spatial database, the second for the related non-spatial attributes. ARC/INFO, for example, manages spatial data in ARC and non-spatial data in INFO. PAMAP GIS uses dBASE III PLUS or any other relational database manager to handle non-spatial data.

Systems in this price bracket can be differentiated most easily on the basis of their original design concept of structuring the spatial data component. ARC/INFO, one of the market leaders in GIS, commenced by utilizing a vector based spatial data structure approach. SPANS is unique in its adoption of a quadtree spatial data structure. PAMAP and TERRASOFT commenced as raster based systems. It would appear that, as software development progresses, vector based systems are adding raster capabilities, and raster systems are developing vector capabilities.

Macintosh GIS include MAP II, MacMap, MacGIS and GIS Tutor. The latter is not a GIS as such but rather a HyperCard Stack which can be used as a teaching tool for those new to GIS terminology and concepts. The first two packages cited are low cost systems designed for teaching beginners how to operate and develop a GIS. Both are raster systems based on the MS-DOS MAP packages developed by Dana Tomlin. MAP II, developed by Micha Pazner and his colleagues at the

University of Manitoba, represents an excellent implementation of the Macintosh interface and includes the use of color on the Mac II series. MacMAP was developed by Glen Jordan at the University of New Brunswick. MacGIS is another similar low cost, raster based GIS developed by Stephen Smith and his colleagues at Cornell University. These systems, and especially MAP II, provide a very sound introduction to the world of GIS on the Macintosh. They represent a viable alternative for those departments and organizations that do not want to commit themselves to one software package costing $10,000 to $30,000. On the other hand there are very few fully fledged commercial GIS systems for the Macintosh. Québec Hydro and Tydac are reputed to be working on a Macintosh version of Tydac's SPANS package which should be available by the time of publication.

GIS software development is a competitive and rapidly advancing industry. Most systems have topological overlay capabilities. Corridor analysis, spatial query and spatial search are common analytical functions. Many systems support, or are in the process of supporting terrain modelling capabilities and some form of interface to remote sensed data. Systems vary in their ability to handle more sophisticated analytical functions, but a new analytical capability supported by one system that proves a popular option will soon be implemented on competitors' products.

Microcomputer based geographical information systems have come a long way in a relatively short time period. Most systems, however, are still not user-friendly and many have a considerable learning curve even for those with an extensive geographical training.

Image Analysis Systems

Some cartographers may be interested in a microcomputer based image analysis system. The process of topographic map creation and land use classification increasingly relies on remotely sensed data. Image analysis systems are clearly of importance to cartography. However, image analysis software tends to concentrate on the manipulation of remotely sensed data. Programs will vary depending on analytical capabilities, processing power and the man-machine interaction interface. Given the emphasis on analytical capabilities and classifications methodologies, a useful discussion of image analysis software is felt to be beyond this chapter although some microcomputer image analysis packages have been included in the database. Those included are APPLEPIPS, DRAGON, EIDETIC, ERDAS, IMAVISION, MICROPIPS and PCIPS. The reader is referred to remote sensing and photogrammetry journals for detailed reviews (see also Wheate 1989).

Other Mapping Software

There exists other microcomputer software with mapping capabilities. This software includes programs written for educational purposes, spatial analysis, geology and specialized map production.

Educational software tends to concentrate on the identification and location of place names or the query of statistical data about places. Spatial analysis software tends to be concerned with problems of transportation and location and allocation. Geological mapping and geological map analysis is an area of microcomputer software development that is rapidly emerging. Some software is being developed for electronic chart production for marine- and road navigation, and the first microcomputer based electronic atlases now are appearing on the market.

With the exception of geological mapping and chart production, mapping is only a minor component of these software packages, and these programs therefore will be discussed only briefly.

Place name and Location Packages

An example of a low resolution graphics package for IBM and compatibles that serves as a place name atlas of the world, offering factual information, is Software concepts computerized atlas for $50 (discussed in more detail below). Another educational global place-name package is GEOGRAPHY for $25.

A number of IBM and compatible oriented educational place packages concentrate on the United States. LOKATE, a $5000 option, specializes in offering the user information on about approximately 60,000 locations in the USA. Another program, USA DISPLAY, operates only on state level data. This package allows for the selective recall of statistical information to be plotted in map format. Aimed at schools as well as businesses and homes, PC USA for $60 will yield information on physical as well as socio-economic and demographic data that can be visually overlain for examination. It would appear that users can modify data using this package.

An attempt at microcomputer based mental map recall has been made with a program called MENTMAP2. However to date, this area of cartography, namely mental mapping and map recall has been little explored in conjunction with microcomputer technology.

An interesting development in microcomputer based mapping is the advent of the digital atlas (see also chapters by Slocum and Egbert, and Raveneau *et al.* in this volume). The ATLAS OF ARKANSAS, produced by the University of Arkansas Press, is a good example. This atlas contains visual slides of over 100 geographical themes, including textual information and a bibliography.

Some very basic electronic atlases have been developed on the Macintosh. Perhaps the most successful are the MacAtlas and ATLAS packages.

Holloway (1986) provides a brief review of these pioneering developments. MacAtlas comes in several volumes and is essentially a boundary file package which allows the user to produce their own map designs using the various "paint" packages which were beginning to proliferate at the time of development.

ATLAS has been marketed by Software Concepts (a division of Rand McNally and Co.), a well-known publisher of traditional atlases. ATLAS was perhaps the first true electronic atlas (see Waters and de Leeuw 1987). It produces a rotating, three dimensional globe on the screen of the Macintosh. Users can move to any spot on the globe by typing in the appropriate latitude and longitude, and can subsequently zoom in on that location by using the scroll bar on the side of the screen. The maps of the various spots on the globe can be customized by cutting them from the atlas and pasting into the MacPaint package. The original version included some basic information for 2500 cities contained in the program's memory. Distances could also be calculated between any one of these cities in land miles, nautical miles or kilometers. Similar graphics are now used by the Knight-Ridder newspaper wire service which uses Macintoshes for all its maps.

On the negative side, this program was about four times the price of a traditional school atlas and the original version only supported monochrome and contained only limited information about the towns located. It was surprisingly parochial in its American bias; information on Calgary indicated it was located in the "State" of Alberta. The original program was heavily copy protected, and had no site licensing agreement or innovative marketing concepts beyond the provision of a cheap demo disk which could be unlocked when the full purchase price was paid. Site licensing has since become more common, but has rarely become inexpensive enough for most schools, although Systat Inc. is making some innovative steps in this direction (see Waters 1989c for a discussion).

Projection Packages

MAPS and WORLD are two microcomputer programs designed for IBM and compatibles that concentrate on map projections. WORLD for $250 supports over 150 named map projections and allows the user to create any number of new ones. The user is given considerable flexibility with respect to parameter specification. One of the drawbacks of WORLD is that it appears to require a dual monitor workstation.

The Map Collection program developed by Frank Gossette at California State (Long Beach), and marketed through his firm MapWare, is an integrated MS-DOS package which performs a variety of map projections, but also has many of the other mapping capabilities discussed earlier, including contour mapping, three-dimensional surfacing, graduated circle, choropleth and line mapping. It is an excellent package for teaching purposes, not only because of its academic orientation, but also because of its very reasonable price of $250 which includes a site license. A recent review is provided by Hepner (1990).

A projection package developed for the ATARI is called MAPS AND LEGENDS: THE CARTOGRAPHER. This package appears to have been designed for the secondary geography teacher and the high school student market.

Applications Software with a Mapping Component

Many application specific, microcomputer programs have been written that include a mapping component. These packages tend to emphasize analytical modules, using mapping capabilities to display results visually, and to allow the user to interact visually with the data when searching for answers. A thorough discussion of all of these programs is beyond this chapter. We will briefly mention two.

The first is TRANSPRO for $995. This is a software package designed for transportation planning and traffic policy. The program allows the user to examine the impact of new roads and shopping malls on traffic patterns. The program will display output in map format.

The second example is LANDTRAK, an expensive microcomputer program written to assist in crime analysis. LANDTRAK allows law enforcement agents to enter crime related data in conjunction with basic information concerning municipalities. A sophisticated set of query capabilities subsequently allows a search of the database for criminal trends and patterns. Locations of crime as well as spatial trends and patterns can be displayed cartographically on the screen and in hard copy format. One cannot help but get the impression that, so far, mapping components accompanying application oriented software have been designed and written by computer scientists, not cartographers. Designing meaningful, effective and cartographically elegant visual display modules for application oriented software packages is an area of cartography that offers tremendous potential for cartographic research and involvement. It is essential that cartographers should seize this opportunity.

Digital Charts

A rapidly growing field of microcomputer mapping concerns the development of electronic charts. Development appears to be occurring simultaneously in three directions. First, there are ongoing efforts to create digital displays of simplified navigational images of chart facsimiles interfaced to navigational instrumentation. Second, there are efforts to interface the paper chart with information and analytical capabilities stored in a microcomputer. Third, there are attempts to create high resolution electronic replicas of official charts to be interfaced with navigational instruments and with added analytical capabilities.

A thorough discussion of the advent of electronic navigation including digital display in aviation, marine and terrain travel is beyond this chapter. The reader is referred to a series of articles by Donaldson (April, May 1989) in *Pacific Yachting* and a paper by Keller (1989) on recent developments in electronic charts.

Discussion

When examining microcomputer product news-releases, one gets the impression that memory capability, increases in processing speed, processing capability and display quality are improving daily. While technology is advancing rapidly, the cost of entering personal computing has dropped considerably and is continuing to drop. Any discussion concerning microcomputers and software written for microcomputers must therefore take into consideration the highly dynamic nature of this technology. Comments valid today are often obsolete a few months later.

The last statement certainly applies to the above discussion of microcomputer mapping software. We anticipate that revised versions of existing mapping software, and many new mapping software products will have been released between writing this chapter and its publication. Nevertheless, we hope that the reader will find this chapter useful since it gives an overview of software available for mapping, representing a window in time.

It may be of some interest to speculate on future trends in microcomputer mapping software. The ultimate goal of cartography is to communicate spatial distributions, and digital cartography to this day has not reached the same quality of communication as analog cartography. Development efforts will therefore aim at improving the quality of digital communication (see also chapter by Slocum and Egbert in this volume). In the process, a number of communication capabilities not handled effectively by analog cartography will be advanced. These will include the capability to communicate interactively with the cartographic database, and the capability to communicate the dimension of time or change. These development efforts will depend to a large extent on technological advances in input and output peripherals and database structures.

Advances in digital cartography will also change the emphasis of cartography away from communication towards cartographic analysis. Considerable efforts therefore will be spent adding analytical capabilities to digital mapping systems. Taylor, in a chapter in this volume, defines these emerging systems as electronic mapping systems.

Digital systems have the potential to handle scale change easily utilizing the principle of zooming. At present a scale change through zooming is not supported by a change in generalization of the cartographic database, that is zooming does not imply a change in symbolization, simplification and omission. Cartographers will therefore have to develop rules, methods, standards and procedures to allow a cartographical database to change as scale changes. Considerable efforts are spent at present developing standards for digital cartography, standards concerning typography, color, symbolization and nomenclature. Where applicable and necessary, these standards will slowly be incorporated into microcomputer mapping products.

Other developments which we expect to take place include the ability to transfer from one graphics file format to another. The Curator program is a step in this direction, although only a limited number of file formats are handled. The ability to manage graphics databases directly is also likely to be made available in the next few years. The Graphidex program provides a rudimentary form of graphics database management in the Macintosh environment. Boundary files and associated data sets will become commonly available in generic formats. Indeed CD ROMs containing U.S. Census Bureau data are already being used in an innovative computer cartography program at California State University, Northridge. Color will become more and more important as the processing power, software and associated peripherals such as scanners, plotters and color laser printers improve in quality and drop in price. Finally, we expect an improvement in user interfaces and an introduction of expert advisors as front ends to the more sophisticated microcomputer mapping software (see chapter by Buttenfield and Mark in this volume). The latter will provide novices and first time users with guidance on producing aesthetic and cartographically pleasing maps. Expert advisors represent a facility which is notable by its absence in the present generation of software which we reviewed.

The speed of the future development of mapping software will depend to a large degree on the size of the market for digital mapping products. The initial market for digital mapping software products has been small when compared to software markets for wordprocessing, spreadsheet or even CAD, with educational institutes and marketing companies being notable for their interest. A small vertical market implies relatively slow development progress. It is anticipated that the market for digital mapping products will increase with the growing popularity of GIS, the introduction of digital mapping displays for navigational instruments, and an increased awareness of mapping as a form of communication and analysis in marketing. It is therefore to be expected that the speed of development will increase over the next few years.

Exciting times are ahead, and keeping up with developments in digital mapping will be a challenge.

TABLE 6.1

Microcomputer Mapping Software

Name	Review	IBM	MAC	CAD	RSP	MAS	GIN	COT	Cost	Company
ALDUS FREE HAND, Version 2.0	MacUser 5(6) 1989, pp. 45–46		y	y	y				495	Aldus, 411 First Ave. S., Seattle, Washington, 98104–2871
APPLEPIPS. Versions 1.0 and 2.0	The Professional Geographer 38(3) 1986		y	y					495	Telseys Group, Columbia, Maryland
ARC/INFO LAB KIT	Brochure	y					y		13,750	ESRI, 44 Upjohn Rd, Don Mills, Ontario, M3B 2W1
ATLAS AMP. Version 1.10	American Cartographer 13(4) 1986	y		y					449	Strategic Locations Planning, San José, California
ATLAS*DRAW	Brochure		y	y					750	Strategic Locations Planning, San José, California
ATLAS*GRAPHICS	Professional Geographer 41(3) 1989, pp. 367–368	y					y	y	450	STSC Inc., 2115 E. Jefferson St., Rockville, MD, 20852
AUTOCAD	Professional Geographer 41(3) 1989, pp. 368–369, American Cartographer 13(3) 1986	y		y					2,850	Autodesk Inc., 2320 Marinship Way, Sausalito, CA, 94965
AUTOSKETCH	Professional Geographer 41 (3) 1989, pp. 369–370, American Cartographer, 14(4) 1987	y		y					99	Autodesk Inc., 2320 Marinship Way, Sausalito, CA, 94965
CANVAS Version 2.0	MacUser 5(4) 1989		y	y					299	Deneba Software, 3305 NW 74th Ave., Miami, FL, 33122
CARO-PC	Brochure	y		y			y		29	CARTO, Inc., Phoenix, Arizona

TABLE 6.1—*continued*

		I	M	C	R	M	G	C		
Name	Review	B	A	A	S	A	I	O	Cost	Company
		M	C	D	D	P	S	N T		
CLARIS CADD	MacUser 5(7) 1989, p. 152		y	y					799	Claris, 440 Clyde Ave., Mountain View, CA, 94043
COMPUGRID	Brochure		y				y		n.a.	Geospatial Systems Ltd., St. Albert, Alberta, T8N 2B2
CONCEPTS COMPUTERIZED ATLAS. Version 01.01.00	American Cartographer 14(4) 1987	y	y						50	Software Concepts Incorporated, Stamford, Connecticut
CONTOUR 81	Brochure		y					y	10	Technical Software Systems, P.O. Box 56, Millburn, NJ, 07041-1008
CONTUR. Version 1.20	The Professional Geographer (37)4 1985	y						y	199	In-Situ, Inc., Laramie, Wyoming
CRICKET PAINT	MacGuide 2(2) 1989, p. 148		y	y					195	Cricket Software, 40 Valley Stream Parkway, Malvern, PA, 19355
CURATOR	MacGuide 1(3) 1988, p. 176a		y						139	Solutions Int., 29 Main St., P.O. Box 989, Montpelier, VT, 05602
DESKDRAW	MacWorld 6(8) 1989, p. 140		y	y					130	Zedcor, Inc., 4500 E. Speedway, #22, Tuscon, AZ, 85712-5305
DESKTOP INFORMATION DISPLAY SYSTEM (GIDS). Version 2.0	American Cartographer 12(2) 1985	y					y		1,500	Sammamish Data System, Inc., Bellevue, Washington

Software	Supplier	Price				Reference
DIAGRAPH	Computer Support Corp., Carrollton, Texas	395	y		y	"Putting Your Business on the Map" PC Magazine 1986
DOCUMAP	Morgan-Fairchild Inc., Seattle, Washington	1,000	y	y	y	The Professional Geographer 38(2) 1986
DRAGON Level 1. Version 1.19	Goldin-Rudahl Systems, Inc., North Amherst, Massachusetts	1,000	y		y	The Professional Geographer 41(1) 1989
DRAWING TABLE	Broderbund Software, Inc., 17 Paul Dr., San Rafael, CA, 94903-2101	130		y	y	MacWorld 6(8) 1989, p. 140
DREAMS	Innovative Data Design	500		y	y	MacUser 5(7) 1989, p. 152
EASYCAD. Version 1.07	Evolution Computing Tempe, Arizona	170	y	y	y	The Professional Geographer 41(1) 1989
ERDAS-PC	Advanced Technology Development Centre	10,000	y	y	y	Brochure
ESLMap	Environmental Sciences Ltd., Sidney, B.C.	700		y	y	Brochure
EXECUVISION	EXECUVISION, Prentice-Hall Inc., Englewood Cliffs, NJ, 07632	295	y		y	Journal of Geography 85(2) 1986, pp. 62–66
FASTCAD. Version 1.20	Evolution Computing Tempe, Arizona	2,295	y		y	The Professional Geographer 41(1) 1989
FIMSYS and FIMGRAF	Fimtech, Inc., Atlanta, Georgia	n.a.	y		y	Brochure
FIRE-ROUTER. Version 4	M. M. Dillon Ltd., Toronto, Ontario	3,500	y		y	The Professional Geographer 40(2) 1988
FREELANCE	Lotus Development Corp., Waltham, Massachusetts	395	y		y	"Putting Your Business on the Map" PC Magazine 1986
FULLPAINT	Ashton-Tate MacDivision, 2393 Teller Rd., Newbury Park, CA, 91320	100		y	y	MacGuide 1(3) 1988, p. 178a
GENERIC CADD	Generic Software, 11911 N. Creek, Parkway S., Bothell, WA, 98011	100		y	y	MacUser 5(7) 1989, p. 152

TABLE 6.1—*continued*

Name	Review	IBM	MAC	CAD	RSP	MAS	GIN	COT	Cost	Company
GEOGRAPHY	Professional Geographer (39)4 1987		y						25	DATASYSTEM, 2301 Churchill Drive, Oxnard, CA 93033
GEOQUERY	MacGuide 1(3) 1988, p. 283a		y	y					350	Odesta Corp., 4084 Commercial Ave., Northbrook, IL, 60062
GEOVIEW	GeoByte 3(2) 1988, pp. 17–21		y	y				y	900	Computer Syotemics, 806 Hill Wood Dr., Austin, TX, 78745
GIS TUTOR	Brochure		y				y		100	GIS World, Inc., P.O. Box 8090, Fort Collins, CO, 80526
GOLDEN SOFTWARE Mapping Package. Version 1.0	The Professional Geographer 37(4) 1985	y						y	199	Golden Software, Golden, Colorado
GRAPHIDEX	MacGuide 1(3) 1988, p. 178a		y						80	Brain Power Inc., 24009 Ventura Boulevard, Calabasas, CA, 91302
GSMAP	GeoByte 3(4) 1988, pp. 36–43		y	y					n.a.	U.S. Geological Survey, 502 National Center, Reston, VA, 22092
GWN-DTM Digital Terrain Model	Brochure	y	y	y				y	18,000	Scientific Publications and Software Center, Washington D.C.
HYPERATLAS	MacUser 4(10) 1988		y	y					99	Micromaps Software, P.O. Box 757, Lambertville, NJ, 08530

Software						Price	Supplier	Reference
IDRISI A Grid Based Geographic Analysis System	y					100	Clark University, Worcester, Massachusetts	Brochure
ILLUSTRATOR 88	y	y	y	y	y	495	Adobe Systems, 1585 Charleston Road, Mountain View, CA, 94039	MacUser 4(10) 1988
IMAVISION	y		y			20,000	Roy Ball Associates, Ottawa, Ontario	Brochure
IN-CAD	y	y				2,495	Infinite Graphics	MacUser 5(7) 1989, p. 158
IRIS GIS	y	y	y			950	IRIS Intl, 4301 Garden City Dr., Landover, MD, 20785	Journal of Geography 85(2) 1986, pp. 62–66
LANDTRAK	y					18,000	Brochure	Brochure
LOKATE						5,000	Distribution by Design Roslyn, New York	The Professional Geographer 37(3)
MACATLAS	y	y				199	Micromaps, Lambertville, New Jersey	The Professional Geographer 40(2) 1988
MACBRAVO	y	y				1,500	Schlumberger	MacUser 5(7) 1989, p. 158
MACCHORO. Versions 1.1 and 1.2	y	y				295	Image Mapping System, Omaha, Nebraska	American Cartographer 14(1) 1987, Professional Geographer 39(4) 1987
MACCONTOUR	y		y			900	SIAL Geoscience, Inc., 969 Route de l'Eglise, Ste. Foy, Quebec	GeoByte 4(3) 198, pp. 48–51
MACDRAFT	y	y				269	Innovative Data Design, 2280 Bates Ave., Concord, CA, 94520	MacUser 2(2) 1986
MACDRAW II	y	y				395	Claris, 440 Clyde Ave., Mountain View, CA, 94043	MacUser 4(11) 1988
MACGEOS II	y	y				725	SIAL Geoscience Inc., 969 Route De l'Eglise, Ste. Foy, Quebec	Geobyte 4(3) 1989, pp. 48–51

TABLE 6.1—*continued*

Name	Review	IBM	MAC	CAD	RSP	MA	GIS	CONT	Cost	Company
MAGGIS	Journal of Geography 88(2) 1989, p. 60		y					y	159	Stephen D. Smith, CLEARS, Cornell University, Ithaca, NY, 14853
MACGRASS	Brochure		y					y	950	Space Remote Sensing Center, Stennis Space Center, MA, 39529
MACMAP	Brochure		y					y	n.a.	Glen Jordan, Center for Resource Information Studies, U. of N.B.
MACPAINT	MacUser 4(7) 1988		y	y					125	Claris, 440 Clyde Ave., Mountain View, CA, 94043
MAP (Map Analysis Package)	Brochure	y	y				y	y	20	Harvard Lab for Computer Graphics and Spatial Analysis, Cambridge
MAP II	Brochure		y					y	100	John Wiley, 605 Third Ave., New York, NY, 10158-0012
MAP-MASTER	"Putting Your Business on the Map" PC Magazine 1986		y					y	395	Decision Resources, Inc., Westport, Connecticut
MAPEDIT Version 1.0	American Cartographer 13(4) 1986, Professional Geographer 39(2) 1987		y						249	Strategic Locations Planning, San José, California
MAPGRAFIX and MAPSTAR	MAC Guide 1(3) 1988, p. 283a		y	y			y	y	8500	COMGrafix, Inc., Clearwater, Florida
MAPIT	The Professional Geographer 38(1) 1986		y				y	y	95	Questionnaire Service Company East Lansing, Michigan

Software	Source			Price	Vendor
MAPMAKER	MACUSER, Jan. 1987	y	y y	349	Select Micro Systems, 40 Triangle Center, Yorktown Heights, NY 10598
MENTMAP2	Professional Geographer (38)4 1986	y	y	n.a.	Lawrence W. Carstensen, Department of Geography, Virginia Tech.
MGMS: PROFESSIONAL CAD FOR MACINTOSH	MacUser 3(11) 1987	y	y	799	Micro CAD/CAM, 5900 Sepulveda Blvd., Suite 340, Van Nuys, CA, 91411
MICROBASED AUTOMATED PROJECTION SYSTEM (M.A.P.S)	Brochure	y		900	RDS System Inc
MICROMAP/DOCUMAP	Brochure	y y	y y y	650	Morgan-Fairchild, Seattle, Washington
MICROPIPS Personal Image Processing System. Version 1.0	American Cartographer 13(1) 1986	y	y	650	The Telesys Group, Inc., Columbia, Maryland
MICROSTATION MAC	MacUser 5(7) 1989, p. 158	y	y	3,300	Intergraph Corporation, 1 Madison Industrial Park, Huntsville, AL
ULTIMAP	The Professional Geographer 38(3) 1986	y	y	495	Planning Data Systems, Philadelphia, Pennsylvania
MUNDOCART/CD	Brochure	y		9,250	Chadwyck-Healey, Inc., Alexandria, Virginia
Mac-GIS	Journal of Geography 88(2) 1989, p. 60	y	y	159	Clears, Cornell University, 464 Hollister Hall, Ithaca, NY, 14853
MacPaint, Version 1.5	The American Cartographer (14)1 1987	y		125	Apple Computer Inc.
NUCOR	Brochure	y	y	n.a.	Nucor Computing Resourcesa Inc., P.O. Box 13520, Kanata, Ontario, K2K1X6

TABLE 6.1—*continued*

Name	Review	IBM	MAC	CAD	RAP	MIS	GON	CONT	Cost	Company
OSU MAP	Brochure	y		y					n.a.	Department of Geography, Ohio State University, Columbus, OH, 43210
PAMAP Geographic Information Systems	Brochure	y					y	y	30,000	PAMAP Graphics Ltd., Victoria, B.C.
PANACEA. Version 1.05	The Professional Geographer 40(4) 1988	y			y	y	y		2,380	Earth Imaging Systems, Inc. Cambridge, Massachusetts
PC MAPICS	The Professional Geographer 40(4) 1988	y				y			760	Mapics Ltd., London, UK
PC-DRAW	The Professional Geographer 38(2) 1986	y		y					395	Micrografx, Inc. Richardson, Texas
PC-USA	Brochure	y							60	Comwell Systems, Inc. Phoenix, Arizona
PCIPS (Personal Computer Image Processing System)	The Professional Geographer 39(1) 1987	y					y		350	IBM Direct Dayton, New Jersey
PCMAP	Journal of Geography 85(2) 1986, pp. 62–66	y			y	y			180	Geo Software, P.O. Box 6042, Macomb, IL, 61455
PIXEL PAINT	MacUser 4(5) 1988		y	y					495	SuperMac Technology, 485 Potrero Ave., Sunnyvale, CA, 94086
POLYMAPS	"Putting Your Business on the Map" PC Magazine 1986	y	y				y		99	Community Research and Information Systems, Ripley, New York
POLYMAPS. Version 2.0	The Professional Geographer 40(4) 1988	y		y					185	Chautauquasoft Ripley, New York

Product	Reference					Price	Company
PRODESIGN II. Version 2.0	The Professional Geographer 39(2), 1987, American Cartographer 13(2) 1986	y			y	299	American Small Business Computers, Inc., Pryor, Oklahoma
PROFESSIONAL REMOTE SENSING IMAGE ACQUISITION AND ANALYSIS SYSTEM	Brochure	y			y	9,950	Eidetic Digital Imaging Ltd., Brentwood Bay, B.C.
QUICKMAP. Version 1.0	The Professional Geographer 38(4) 1986	y			y	295	Sammamish Data Systems, Inc., Bellevue, Washington
RANDMAP. Version 1.1	American Cartographer	y			y	495	Rand MaNally & Company, Chicago, Illinois
RANDMAP. Version 1.2	The Professional Geographer 39(2) 1987, PC Mag. Sept 30, 86	y			y	495	Rand McNally & Company, Chicago, Illinois
ROOTS	Brochure	y				20	Harvard Lab for Computer Graphics and Spatial Annalysis, Cambridge
SALADIN	Brochure	y			y y y	n.a.	Research Centre for Urban and Regional Planning, Delft, Netherland
SAS/Graph	American Cartographer 14(2) 1987	y				n.a.	SAS Institute
SPSS GRAPHICS	American Cartographer 14(2) 1987	y			y	8,000	SPSS Inc., Chicago, Illinois
STATE DATA SYSTEM	The Professional Geographer 38(2) 1986	y	y		y	20	SSMR Software Department, North Carolina State University, Raleigh
STATMAP	Journal of Geography 85(2) 1986, pp. 62–66	y				995	Rand McNally, Infomap, 8255 N. Central Pk. Ave., Skokie, IL, 60076
SUPERPAINT	MacUser 3(2) 1987			y	y	150	Silicon Beach Software, 9580 Black Mountain Road, San Diego, CA

TABLE 6.1—*continued*

Name	Review	IBM	MAC	CAD	R S	MAP	GIS	COS	CONT	Cost	Company
SURFACE GRAPH	Brochure	y							y	25	Jeff Amfahr, 750 W. Ave. J-8, Lancaster, CA, 93534
SYSTAT WITH SYSGRAPH	PC Magazine 8(5) 1989, pp. 94–162	y	y				y		y	795	Systat Inc., 1800 Sherman Ave., Evanston, IL, 60201
T-Mapper, TerraPak Software	Brochure	y					c		r	n.a.	Terra-Mar Resource Information Services, Inc., Mountain View, Cal.
TERRASOFT	Brochure	y							y	30,000	Digital Resources Systems, 103–238 Franklyn, Nanaimo, B.C., V9R 2X4
THE GOLDEN GRAPHICS SYSTEM. Version 2.0	American Cartographer 13(3) 1986	y							y	200	Golden Software, Inc., Golden, Colorado
THE MAP COLLECTION	Brochure, AAG Microcomputer Specialty Group Newsletter 5(2) 1989	y					y		y	150	Frank Gosette, Mapware, P.O. Box 50168, Long Beach, CA, 90815
THUNDERSCAN	The Professional Geographer 39(4) 1987	y					y			229	Thunderware, Inc., Orinda, California
TRANSPRO. Version 2.20	The Professional Geographer 40(2) 1988	y					y			995	Transware Systems Irvine, California
TRIMAP	Brochure	y							y	n.a.	Contoursoft Corp., Edmonton, Alberta
TURBOCAD	Professional Geographer 41(2) 1989, pp. 234–235	y		y						100	Pink Software Development, Milan Systems America Inc., Atlanta, GA

Software	Reference					Price	Vendor
TYDAC SPATIAL ANALYSIS SYSTEM (SPANS)	Brochure	y	y	y	y	3,000	Tydac Technologies, Arlington, Virginia
UNLOCKING THE MAP CODE, TIME AND SEASON, CHOICE AND CHANCE	The Professional Geographer 37(1) 1985	y				110	Rand McNally, Educational Publishing Software Division
URBAN DATA MANAGEMENT SOFTWARE (UDMS)	The Professional Geographer 39(1) 1987, 40(4) 1988	y	y	y		n.a.	United Nations Centre for Human Settlements (HABITAT) Nairobi, Kenya
USA DISPLAY	The Professional Geographer 38(4) 1986	y		y		n.a.	Instant Recall, Bethesda, Maryland
VCN Concorde	"Putting Your Business on the Map" PC Magazine 1986	y	y	y		695	Visual Communications Network, Cambridge, Massachusetts
VERSACAD	MacUser 4(7) 1988	y	y	y		1,995	VersaCad, 2124 Main Street, Huntington Beach, CA, 92648
VGS 300 PLUS	Brochure	y			y	n.a.	Hunter and Associates, Mississauga, Ontario
WINDOWS/ON THE WORLD*GEOdisc U.S. ATLAS	GEOBYTE 3(2) 1988, pp. 87–88	y		y		595	Geovision, Inc., Norcross, Georgia
WORLD	The Professional Geographer 40(4) 1988	y				250	University of Minnesota, Minneapolis, Minnesota

References

Abernathy, E. and P. Weiss (1989) "Gray expectations", *MacUser*, Vol. 5(6), pp. 170–193.

Beale, S. and J. Cavuoto (1989) *The Scanner Book*, Micro Publishing Press, Torrance, California.

Bortman, H. (1989) "Scanning and the color horizon", *MacUser*, Vol. 5(6), pp. 90–117.

Carswell, R. J. B., G. J. A. de Leeuw and N. M. Waters (eds.) (1987) "Atlases for schools: design principles and curriculum perspectives", *Cartographica*, Vol. 24, No. 1, Monograph 36, 159 pp.

Domier, J. E. J. (1986) "Reviews of geographic software: MAPIT", *The Professional Geographer*, Vol. 38(1), pp. 101–102.

Donaldson, Sven (1989) "Chart evolution", *Pacific Yachting*, April, pp. 20–22.

Donaldson, Sven (1989) "Electronic charts", *Pacific Yachting*, May, pp. 24–26.

Dutton-Marion, K. (1986) "Macintosh graphics hardware/software: vision systems", *Wrapple*, Vol. 15, pp. 6–17.

Eastman, J. R. (1988) "IDRISI: a geographical analysis system for research applications", *The Operational Geographer*, No. 15, pp. 17–21.

Fisher, P. F. (1988) "Reviews of geographic software: PC MAPICS", *The Professional Geographer*, 40(4), pp. 466–467.

Goodchild, M. F. (1985) "Reviews of geographic software: CONTUR, version 1.20", *The Professional Geographer*, Vol. 37(4), pp. 493–494.

Hepner, G. F. (1990) "Reviews of geographic software: the map collection educational version 1.25", *The Professional Geographer*, Vol. 42(1), p. 125.

Holloway, M. (1986) "Round the world with Mac", *The MACazine*, Vol. 3(2), pp. 73–73.

Keller, C. P. (1989) "Digital charts for navigation", *Northwest Cartographer*, Vol. 29(3), pp. 6–8.

Lewis, L. T. (1988) "Reviews of geographic software: POLYMAPS", *The Professional Geographer*, Vol. 40(4), pp. 467–468.

Lipsey, M. (1987) "Software reviews: AutoSketch. version 1.0", *The American Cartographer*, Vol. 14(4), pp. 367–370.

Mak, K. (1989) "Classed and unclassed map communication", Master's Thesis, Department of Geography, University of Calgary.

McMaster, R. B. (1989) "Reviews of geographic software: EASYCAD", *The Professional Geographer*, Vol. 41(1), pp. 88–89.

Myers, R. (1985) *"Microcomputer Graphics"*, Addison-Wesley, Reading, Massachusetts.

Peltz, D. L. (1988) "3-D in perspective", *MacWorld*, Vol. 5(12), pp. 108–119.

Peltz, D. L. (1989) "Hands-on CAD", *MacUser*, Vol. 5(7), pp. 152–167.

Rudnicki, R. (1987) "Reviews of geographic software: PRODESIGN II", *The Professional Geographer*, Vol. 39(2), pp. 232–233.

Sechrist, R. P. (1986) "Reviews of geographic software: PC-DRAW", *The Professional Geographer*, Vol. 38(2), p. 193.

Snaden, J. N. (1986) "Reviews of geographic software: MULTIMAP", *The Professional Geographer*, Vol. 38(3), pp. 293–294.

Walker, G. (1985) "The microcomputer in university cartographic teaching", Ch. 7 in *Education and Training in Contemporary Cartography*, edited by D. R. F. Taylor, Wiley, New York.

Waters, N. M. (1989a) "Geography, microcomputers and the use of color", *The Operational Geographer*, Vol. 7(3).

Waters, N. M. (1989b) "What's next? A look back and forward to the future of computing in geography", *The Operational Geographer*, Vol. 7(1), pp. 30–34.

Waters, N. M. (1989c) "Statistical analysis and forecasting packages for microcomputers", *The Operational Geographer*, Vol. 7(2), pp. 30–33.

Waters, N. M. and G. J. A. de Leeuw (1987) "Computer atlases to complement printed atlases", in Carswell, R. J. B., G. J. A. de Leeuw, and N. M. Waters (eds.), op. cit., pp. 118–133.

Wheate, R. (1989) "Image processing with microcomputers", *The Micro Byte*, Vol. 2(3), pp. 1–5.

CHAPTER 7

Expert Systems in Cartographic Design

BARBARA P. BUTTENFIELD and DAVID M. MARK

National Center for Geographic Information and Analysis
Department of Geography
State University of New York at Buffalo
Buffalo, New York, 14260, USA

Introduction

Map making is a long-standing human endeavor, and was an early area for the application of computers. However, computer cartography thus far has been dominated by automated drafting. Automated map making requires a large degree of human interaction in order to produce products of acceptable quality. This is because many aspects of the cartographic process require human judgement, and the application of aesthetic principles. Human cartographers can learn to produce adequate maps after only a few hundred hours of instruction. Instruction commonly involves large amounts of hands-on practical work, followed by expert critique and repeated practice—essentially an apprenticeship mode of learning.

From the above description, it is evident that map design is a good example of the type of problem that expert systems are often used to solve (see definitions and review, below); in fact, an expert systems approach may be essential if computers are to produce high quality maps with little human intervention (Robinson and Jackson 1985). Computer production of maps is essential in a number of real-time mapping tasks, especially in support of navigation and in conjunction with geographic information systems (GIS); Frank *et al.* (1987, p.p. 266–268) identify cartographic experts as crucial to the applications of GIS for data analysis and depiction.

This chapter emphasizes application of expert systems for reference and locational maps. The broader question of expert systems for the design of more abstract maps, such as statistical maps, has seldom been addressed in the literature. It seems clear that an expert system for design of reference maps, locational maps, and some navigation aids may be more feasible in the short term than a system to generate locational maps as well as purely thematic and analytical graphics. In part this is because the symbology of locational maps is more standardized and thus design constraints may be more readily identified. The specificity of map purpose for locational maps also places finite bounds on the range of appropriate symbology, without limiting map projection, scale, use of color, and level of generalization.

The design of statistical maps involves a more complex and intuitively guided decision process. Design of maps for analytical use might include graphics to display output of iterative models, or design of electronic "throw-away" maps used to formulate hypotheses, or in sensitivity analyses, to cite a few examples. Often these exploratory products are intended for short-term use by a single researcher, and the map form, content and visual hierarchy may not be well defined. While an expert system approach may provide sound principles for setting initial graphic defaults, the guidance of the user and the particular research domain will compound the difficulties of automating map design decisions.

The amplified intelligence approach argued for by Weibel and Buttenfield (1988) incorporates elements of an expert system with the flexibility of human interaction. This type of approach may prove simpler than a pure artificial intelligence solution for some aspects of map design, by removing much of the rule base from the system and letting the (proficient) user guide many design decisions directly. The need for interactive visualization tools also has been discussed by Ganter (1988).

This is not to imply that an expert system to accommodate analytical graphics is impossible, but rather that the intuitive nature of map design is more readily formalized for illustrative than for analytic graphics, and such formalization is prerequisite to the generation of a rule base. (As described below, the rule base in an expert system initiates the flow of system operation.) Where formalized descriptions of rules are not common, alternative approaches may substitute for, or perhaps augment the expert systems solution. For illustrative mapping, and particularly locational mapping, design principles may be readily formalized. For this reason the following discussion will primarily emphasize expert systems for design of locational and navigational maps and charts.

The purpose of this chapter is to present design criteria for a cartographic expert system that focuses on the problems of map execution, including compilation and production of base features and thematic overlays of locational information. Expert systems and artificial intelligence for other cartographic processes, such as digital scanning, or knowledge-based geographic information systems will not be considered. The discussion remains on a conceptual level. The components of a system and the required functionalities are identified, and the relationships between the two discussed. Progress to date on each phase of the cartographic process will be assessed through a review of the literature. First an overview of expert systems design is given.

Expert Systems: Terms and Definitions

Like most popular topics, expert systems have been subject to enthusiasm, "over-selling", skepticism, mistaken ideas, and often-confusing or contradictory terminology. (Artificial Intelligence, or AI, has perhaps suffered even more severely from such problems; AI includes expert systems, cognitive science, and a number of related fields, and will not be further discussed in this chapter.) It is argued that it is important to present clear definitions of what we mean when we use various terms and concepts in this chapter. The terms and definitions used are, to the best of our knowledge, in general accord with the use of these in computer science, and by authorities on expert systems in cartography. For convenience and consistency, we have decided to present definitions from a book introducing "rule-based programming" by Brownston *et al.* (1985). The authors promote a programming language called OPS5: however, our use of their definitions of concepts is not intended to be an endorsement of OPS5 as well.

> "Expert System. A computer program, often written in a production-system language, that has expertise in a narrow domain" (Brownston *et al.* 1985, p. 447).

A production-system language refers to a specialized compiler or symbolic interpreter for implementing an expert system in a particular way. Prolog and LISP are examples of production-system languages. Two other terms used in the definition of expert system require additional clarification:

> "Expertise. Proficiency in a specialized domain. An expert system is said to have expertise in its domain if its performance is comparable to that of a human with five to ten years of training and experience in the domain" (Brownston *et al.* 1985, p. 447).

> "Domain. . . . In expert systems, a field of knowledge or class of tasks" (Brownston *et al.* 1985, p. 446).

Obviously, the domain of interest is cartography; the long-term goal of a research initiative in cartographic expert systems in to produce maps comparable to those currently produced by professional cartographers with five or more years of mapmaking experience.

The next term of importance is knowledge:

> "Knowledge. Any information that can be represented as either declarative knowledge or procedural knowledge, e.g. in the form of rules, entries in data memory or another database, or control strategies. Knowledge may be specific to a task domain or general enough to be independent of all domains" (Brownston *et al.* 1985, p. 448).

The term "knowledge-based system" has appeared in the geographic literature (for example, see Peuquet 1984; Smith 1984; Smith and Pazner 1984). "Knowledge-based system" is almost a synonym for "expert system". However, Brownston *et al.* (1985, p. 449) cast "knowledge-based system" as a more general term, with "expert system" being one intended to capture human expertise.

The distinction between "declarative knowledge" and "procedural knowledge" also is important:

"Declarative Knowledge. Knowledge that can be retrieved and stored but cannot be immediately executed; to be effective, it must be interpreted by procedural knowledge" (Brownston *et al.* 1985, p. 445).

"Procedural knowledge. Knowledge that can be immediately executed using declarative knowledge as data but that may not be examined" (Brownston *et al.* 1985, p. 452).

In a cartographic expert system, declarative knowledge could involve the identification of a line feature making up a political boundary as a ROAD (cultural feature) or a RIVER (naturally-occurring feature). Procedural knowledge might involve setting an appropriate tolerance value for a simplification algorithm. Clearly, both types of knowledge would be required for appropriate generalization of the political boundary. In this case, the procedural knowledge (the rule by which the tolerance value is set) follows from the declarative knowledge (that different feature types require different tolerance values). However, this will not be the case for all situations in cartographic design.

The term "rule" has been mentioned above. Rules lie at the heart of many expert systems:

"Rules. A unit of representation that specifies a relationship between situation and action. Rules are ordered pairs that consist of left-hand side and a right-hand side. ... In production systems, rules are the units of production memory and are used to encode procedural knowledge. A rule is also called a production" (Brownston *et al.* 1985, pp. 453–454).

A rule's ordered pair is often expressed as an If-Then statement. The left-hand side referred to above presents the antecedent condition (the "If" statement) required for the rule to be enabled, and the right-hand side (the "Then" statement) presents the consequent operation to perform.

To bring together the various pieces of an expert system, consider the concept of a production system, which coordinates the knowledge and activity within the system.

"Production system. An architecture for problem-solving that employs a set of rules (stored in production memory), a global database (stored in data memory), and an inference engine that performs the recognize-act cycle of match, conflict resolution, and rule firing" (Brownston *et al.* 1985, p. 452).

Many authorities assert that the essence of a "true" expert system is determined when these three components are distinguished as individual components (cf. Frank *et al.* 1987). The components and operations of a hypothetical production system are illustrated in Fig. 7.1.

Both the data base and the rule base are contained in memory or as mass storage, as shown above. The data base contains coordinates, graphic primitives for typesize, font and style, for line types, shading patterns, and color look-up tables. The rule base contains both procedural and declarative knowledge used to access the database, and may be expanded automatically or during "training sessions", in response to good and bad examples of map design situations.

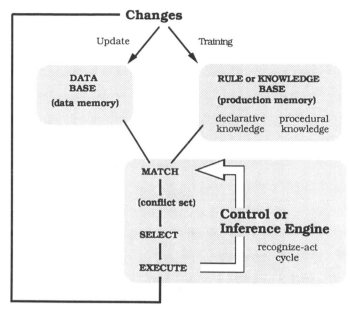

FIG. 7.1. Components of an expert system (adapted from Brownston *et al.* 1985, p. 6).

To take a simple example, procedural rules might guide the selection of a particular color progression for hypsometric tints. The digital values for the color progression would be stored as declarative knowledge, and several different hypsometric progressions might be available in the rule base, perhaps depending upon the selection of an output device. The data base could be accessed to convert the values from the appropriate color look-up table.

The actual flow of control in the production system is directed by the inference engine.

> "Inference engine. [1] The portion of a production-system language that performs inferences by executing the recognize-act cycle. [2] The portion of an application program that performs tasks related to inferring new knowledge in the task domain, as opposed to those portions that perform tasks such as control, input-output, and optimization" (Brownston *et al.* 1985, p. 448).

One problem in expert systems architecture is that many rules may be applicable at the same time, and the execution of any one of these may influence the applicability of the others. Many production systems solve this problem by use of the "recognize-act cycle" mentioned above. First, all rules are matched against the data in the data base. Any rule for which the antecedents (left-hand side) are satisfied is said to be "triggered", and is added to the "conflict set". If the conflict set contains more than one triggered rule, then the inference engine must perform some consistent form of conflict resolution, to select one rule to be "fired". "Firing" a rule means to execute the consequent (right-hand side) of the If-Then statement contained in the rule. After the selected rule is executed, the recognize-act cycle is repeated, until on an iteration the conflict set is empty and the program terminates.

Lastly in this surfeit of terms and definitions, the concept of explanation must be presented, in which a production system recounts the rules and resolved conflicts in the order of their selection and execution. Brachman *et al.* (1983, p. 32) list explanation as a fundamental property of an expert system. "Some researchers believe a capability for explanation is one of the most important features that an expert system can have" (Brachman *et al.* 1983, p. 42).

> Explanation. The process of describing how an expert system reached its conclusions or why it asked particular questions of a user. Explanations may be used to justify decisions or problem-solving strategies, or to teach these strategies to the user (Brownston *et al.* 1985, p. 447).

Cartographic Expert Systems

Recently, Fisher and Mackaness (1987) asked "Are cartographic expert systems possible?" They noted that none of the published accounts of cartographic "expert systems" were able to explain the reasons that decisions were reached, and thus failed to qualify as true expert systems. Fisher and Mackaness (1987, p. 530) expressed the opinion that true expert systems for cartography are possible, and urged the cartographic community to address the problem of "a lack of a systematized and accepted methodology for cartographic assessment" that could be used to evaluate the performance of cartographic expert systems.

Muller, Johnson, and Vanzella (1986) designed and implemented a rule-based system to select a particular thematic symbolization scheme given a categorical description of map user requirements. Requirement categories included type of data, scale of measurement, accuracy requirements, number of variables, and intended map function. Categories available for symbol selection focused on map projection, symbol type (graduated circle, choropleth, contour, etc.), and selection of graphic imposition and graphic primitives. Provisions for "training" and automatic rule modification were demonstrated to show the system capabilities for learning and the efficacy of initializing the rule base with expert knowledge. However, the system included no provision to "explain" its decisions. Furthermore, map execution by automated methods was not addressed in the research.

A full cartographic expert system would be capable of producing, without human intervention, maps of all types, ranging from graphics for rapid on-line GIS display to high-quality printed wall and atlas maps. The system would produce maps for any data set and situation. Maps made by the system should be as effective as the maps produced by professional cartographers, and base-map information would be derived from a single data base. The cartographic expert system would allow the user to specify scales, projections, colors, symbols, and other map elements, but would make good decisions about defaults for any of these if the user chose not to specify them. Ideally, such a system should even make an appropriate choice of the type of cartographic representation (choropleth, isarithm, proportional symbols, cartogram, etc.) if the user did not wish to make such a decision. (This type of decision may or may not be of primary importance in the design of reference and locational maps, but it is nonetheless an appropriate long-term goal for implementation in a cartographic expert system.)

Clearly, the development of such a cartographic expert system is a monumental task. In fact, it may be a problem which can be effectively addressed only by a multi-investigator team over a period of several to many years.

Map Design

Map design includes the entire process by which maps are made. Map design involves abstraction from the real world, and the encoding of real world detail for map representation. Availability of space on the map puts constraints on the selection of objects that can be represented, and on the degree of abstraction required. Graphic limits imposed by media of reproduction, and considerations of aesthetics and clarity, further constrain the map-making process. Recently published textbooks by Cuff and Mattson (1982) and by Dent (1985) are key sources of information on map design principles. The map design process may be divided into three interrelated components: *generalization, symbolization,* and *production* (Fig. 7.2).

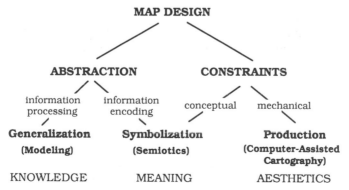

FIG. 7.2. The map design process (from Mark and Buttenfield 1988, p. 417).

In a cartographic expert system, these major components, and the various problems within them (such as point-feature labelling, line generalization, displacement, etc.) would, in expert systems terminology, be represented as contexts within the system:

> Context. A state in a problem-solving process. In a production system, the context may be represented by a special working memory element, which is often called a context element, control element, or goal. Often conceptually isolatable tasks that must be performed by production systems may be partitioned into subtasks that, once initiated, are expected to run to completion. The rules that constitute this task each have condition elements that must match the associated context element (Brownston *et al.* 1985, p. 444).

The fact that the subtasks involved in making a map can be relatively easily isolated may be attested to by the large number of published articles reporting studies of particular modules of design. For example, the choice of map projection may be isolated from the choice of symbology (although the reverse is not always true). Symbol scaling is often treated in isolation of map layout. Legend design and inset composition are most often determined without respect to type selection and name placement. And so forth. This modularity confirms the adaptability of map design and production to an expert systems approach.

Generalization

Relating observations of geographic space to models of geographic phenomena converts data to knowledge. Once a model of the phenomenon has been developed, it is possible to generalize that model. Cartographic generalization may be considered to be an exercise in applied geography (Pannekoek 1962); for example, generalization of contours, coastlines, streams, and topographic surfaces should incorporate knowledge of landforms and geomorphic process. Furthermore, the cartographic character of many features is scale-dependent (Buttenfield 1989), and the use of a single data base to support mapping at a variety of scales presents many research challenged (Beard 1987b).

Several sub-processes are involved in cartographic generalization:

1. *Simplification*. Simplification involves three types of operations at a simple geometric level. *Reduction* of the number of coordinates used to represent a line is often based on tolerance thresholds; phenomenon models are not essential to this sub-process. McMaster's (1987) work on line generalization has until recently emphasized line reduction algorithms.

Also included is *selection*, that requires the identification of features and patterns and their inter-relationships. Töpfer and Pillewizer (1966) discussed heuristics for determining how many features of some class should be shown when a smaller-scale map is compiled from larger-scale sources; however, their model does not address which features should be selected. Priority for feature selection is assessed within the context of the phenomenon being represented, the purpose of the map, and the intended map reader. Buttenfield (1987) proposed a method for generating rules by which features may be identified automatically, but did not implement the method in a working knowledge base.

A third operation is used to *reposition* coordinates to eliminate insignificant detail which may be method produced. For example, McMaster (1987) suggested using a weighted moving average to remove very high frequency crenulations from manually digitized lines. Monmonier (1988) reported initial algorithmic solutions for feature repositioning to improve visual clarity. His solution incorporated interpolation with displacement, which should emphasize to the reader that generalization operations do not fit easily into a discrete typology; that is, one class of operations (e.g. displacement is a part of simplification) often blends into another (e.g. interpolation is a part of enhancement, discussed below).

2. *Classification.* Classification assigns individual objects to categories, and includes the *aggregation* of nearby objects assigned to a common category or to closely related categories by boundary dissolution (Beard, 1987b, calls this operation "coarsening"). *Partitioning* is required for categorizing objects, introducing class breaks for metric attributes, and to select a contour interval in mapping terrain or volumetric data. *Overlay* is a refinement of map information, and is applied in compositing map variables: manual application produces a dasymetric map, and machine-produced weighted models are commonly used to compile land suitability maps, soils maps, and other types of discrete map models in GIS software.

3. *Enhancement.* Enhancement is the purposeful and controlled introduction of information to augment or emphasize structures already present in the data set. "Introduction" should not be misinterpreted here: '. . . enhancement is not the bringing in of unrelated information so much as the amplification of information which is already present" (Buttenfield 1984, p. 8). This may involve *smoothing* operations, as in filtering tasks, *interpolation,* as in threading of contours and isolines across a digital matrix, or *reconstruction* and *generation* of features (such as in 'fractalization"; cf. Dutton 1981). In every case, enhancement involves insertion of data, and this is its primary distinction from simplification.

Brassel and Weibel (1987, p. 125) have distinguished two types of map generalization. The label "statistical generalization" is applied to "spatial modeling for the purposes of data compaction, spatial analysis, etc.", whereas "cartographic generalization" is the label given to "spatial modeling for communication". It is argued that statistical generalization should be viewed as a heuristic, to be avoided after initial data collection and cleaning unless computing power is insufficient to handle a complete data set. In contrast, cartographic generalization, as described by Brassel and Weibel, will take on great importance with attempts to support several scales of analysis and display from a single data base (cf. Beard 1987a, 1987b).

The research on cartographic generalization cited above has utilized standard programming techniques (including recursion) and basic principles from geometry and statistics. Perceptual research has confirmed that some of these standard algorithms replicate point selection by human operators (Marino 1979; White 1985). However, in traditional cartography, the generalization process does not focus on point selection, but rather on preserving the essence of the phenomenon. Development of a typology of generalization operations in a digital context has only recently begun to be reported (McMaster and Shea 1988; Shea and McMaster 1989).

As noted in the introduction to this section, generalization is both a computational and a cognitive process, relating geographic models to observations. A knowledge-based approach to cartographic generalization must first recognize what kind of phenomenon was being generalized, and then use phenomenon-specific rules and procedures (see Mark 1989). Rule-based cartographic line generalization using an expert systems approach has not yet reached the published literature (but see Zhang, Haihong and Xiaochin 1988), if indeed it has seriously been attempted. Certainly this is an important area for future research.

Symbolization

Symbolization is concerned with choosing the symbols, line weights, marks, colors, etc., used to represent map information. As illustrated in Fig. 7.2, there are two separate components of the map design process involved in symbolization. The first encodes information, and is an abstraction process related to generalization. The second involves conceptual constraints. This component relates directly to aspects of production, and focuses on the manifestation of the graphic representation. Both components have a theoretical basis in the field of semiotics. Semiotics addresses the relations among symbols, signs, and meanings, and forms an appropriate basis for cartographic symbolization (Bertin 1983; Child 1984; Schlictmann 1985).

1. *Information encoding.* Information encoding continues the abstraction of map information. The cartographer takes the generalized model of the phenomena to be mapped, and encodes a set of symbols to represent elements and relations of the model. It is highly desirable that the symbols chosen have a "natural" (or at least readily learned) relation to the objects and categories being symbolized. For example, Child (1984) discussed the selection of fish symbols to represent hatcheries on the Washington State highway map. Red fish identified salmon hatcheries, and green fish identified trout hatcheries. Additional meaning was imparted by the twofold orientation of the symbols, with salmon facing upstream to imply their geographic orientation in returning to their spawning grounds.

Encoding statistical information in thematic map-making involves choosing isarithms, choroplethic shading, cartograms, graduated symbols, or some alternative scheme to represent data. Here the natural relation of the symbols to the objects and categories being symbolized may be thought of in statistical or at least data-driven terms. That is, the application of choroplethic shading implies enumerated data, whereas isarithmic symbols may be chosen to depict an interpolated continuous data surface. The consequence of choosing the choroplethic representation is that data must be areally standardized to eliminate the visual bias of map size on perception of choroplethic tones; and areal standardization further abstracts the data.

2. *Conceptual constraints*. Conceptual constraints operate directly upon the graphic depiction, in contrast to the focus in encoding (above) on geographic phenomena. The cartographer strives to maintain a natural semiotic relation between the manifest symbol (that is, the icon) and the visual expectations of the map reader. For example, the colors red and blue carry strong connotations about temperature, larger graduated symbols imply a higher variable value than do smaller symbols appearing on the same map, and without some indication to the contrary, the convention among American map readers is to presume that intermediate scale locational maps are oriented with North at the top of the page. For example, Patton and Crawford (1966) demonstrated the semiotic confusion that may arise when untrained map readers are presented with ambiguous conceptual constraints. Maps of unfamiliar regions showing hypsometric tints were mistaken for displays of vegetation (green tints thought to be woodlands, blue tints thought to be water), temperature, and other phenomena unrelated to terrain representation. Comprehension of map topic sometimes varied with the choice of color progressions making up the hypsometric tints.

On a different level, the limitations of the human visual processing system constrain symbolization, and a large body of cartographic research has focused on development of psychophysical compensations for estimation of size (e.g. Flannery 1971), value (e.g. Williams 1960; Kimerling 1975), visual texture (Smith 1987) and color (e.g. Olson 1987; Heyn 1981). Studies of typesize and style have been reported (Shortridge 1979). Still other research focuses upon conceptual constraints for size and density of dot map symbols (McKay 1949), and three-dimensional representation of classed and interpolated data (Jenks 1963).

3. *Other (situational) constraints*. Cartographic tradition also plays a major role in symbolization, and this is particularly true for locational and navigational maps. When the time available for map reading and comprehension is minimized, as in using a road map while driving a car, map design should depart from traditional symbol choices (such as black cross-hatched lines for railways) only when other considerations make such departures unavoidable. This provides an example of how a design constraint might be formalized, as the preceding constraint could be easily specified as an "if-then" procedural item in a rule base. Attempts to objectify the overall visual

effect of a map have been attempted with varying degrees of success (Petchenik 1974; Gilmartin 1978). This line of research must be pursued and refined to facilitate mechanisms for automatic rule generation by an expert system.

Symbolization must also be applied under constraints of the reproduction and or display technology to be employed. For example, value progressions varying from light to dark often used to be reversed when transferring a map from printed page to computer CRT screen, simply to accommodate the transition from reflected light to transmitted light. In fact, many of the design guidelines established in the research described in section 2 above have not been verified in an electronic map reading situation, and there is evidence for at least some symbols that perceptual continua vary depending on the type of technology. Perceptual grayscales established for photomechanically produced screens have been found to differ substantially from the perceptual scale established for laserprinter-produced screens at the same measured texture (Leonard and Buttenfield 1989), and differences can also be expected for graytones and color progressions produced on CRTs (Heyn 1981).

To summarize, map symbolization provides a bridge between the generalization and map production phases of map design, and incorporates aspects of both abstraction and conceptual constraints in its operation. Symbolization should be based in part on semiotic considerations, and also on perceptual and cognitive findings as well as on principles of computational vision. Muller *et al.*'s (1986) research provides an appropriate avenue by which to address symbolization issues in the context of a cartographic expert system.

Production

Map production involves the actual construction of the graphic image, that is, "making the map". In some ways, the application of expert systems may be centralized in map production, where resolution of alternatives is of key importance. Given that display space is limited, interaction among map elements may lead to spatial conflicts; if these conflicts cannot be resolved through the kinds of adjustments discussed in this section, the map design process must return to the generalization phase. Considerations of graphic design such as clarity and aesthetics are important in map production. Tufte (1983) provides valuable insights into the theory of data graphics in general, and includes many example of statistical mapping in his arguments.

Map elements may be divided into two broad categories: geographic objects, which include point, line, and area features and their symbols, are those elements which have geographical locations; and non-geographic objects, such as titles, legends, insets, north-arrows, scale bars, etc., whose placements are not constrained by geographic reality. Labels for geographic objects represent an intermediate category of map elements, since their possible locations are variable yet spatially constrained.

1. *Plotting*. Plotting of geographic objects does not appear to be an expert systems problem, since it involves the application of deterministic mathematical functions (map projections followed by affine transformations) to the geographic coordinates of the features. Difficulties may arise when spatial conflicts occur at final map scale. However, rules to select the most appropriate map projection would be part of a full CES, and would involve a knowledge base of projection characteristics and properties (Nyerges and Jankowski 1989).

2. *Layout*. The layout of map objects includes decisions about general composition (where does the title go relative to the map, should the legend design be horizontal or vertical, etc.) and also some scaling and clipping operations. Of particular importance is the issue of when and where to incorporate map insets for clarity; a system for map production at the US Bureau of the census includes a rule-based approach to insetting (Beard, Broome, and Martinez 1987; Martinez 1989).

Robinson *et al.* (1984, pp. 147–148) address this issue very concisely, promoting the use of hand-drawn thumbnail sketches to help the cartographer visualize a balanced layout of map elements (including title, legend, map body and inset material). Their approach would prove difficult to automate. Dent (1985) expands upon the topic of composition and layout, enumerating more specific rules for preservation of visual balance of map objects. This approach would be less difficult to implement in an expert system, and might be most efficient to the design process when implemented in an amplified intelligence mode (Weibel and Buttenfield 1988). The interactive drawing and painting software currently available on many microcomputers (see the chapter by Keller and Waters) provides ready manipulation of bitmap and object-oriented graphics, and many of the packages allow some conversion between these formats. An amplified intelligence module could be developed to keep inventory of object relationships established by the cartographer, without requiring the formal expression of those relationships outside the context of a particular map or mapmaker.

3. *Displacement*. Most often, this operation is performed during map production, and refers to the process whereby the map locations of geographic objects are adjusted from their "true" (i.e. archived) locations in order to preserve spatial relations, to provide clarity, or merely to fit other objects onto the map. The rules associated with displacement will be heuristic. For example, point symbols must be "near" the map locations of the point features they represent. Displacement also must recognize relations among point, line, and area features, and preserve those relations during layout and composition. A classic example is the set of lines (river, railroads, highways. etc) running along the floor of a narrow river valley

(Pannekoek 1962). In order to preserve spatial relations, a map design program must establish all the spatial relations prior to displacement.

Also, it would be desirable to determine whether these are functional (causal) relations, which must be preserved, or merely proximity relations, that may be treated more flexibly. Nickerson and Freeman (1986) have applied an expert systems approach to the placement of cartographic lines, and Mackaness and Fisher (1987) propose a system for placing all types of cartographic point features.

4. *Label placement.* An expert systems approach has been applied more often and more successfully to the placement of feature labels than to other aspects of map design. The term "expert systems *approach*" is used because the label-placement programs described in the literature generally are not expert systems in the strict sense as defined above and in the computer science literature (with the possible exception of Jones and Cook 1989).

Label placement is a process with a level of geometric freedom intermediate between that of displacement of geographic objects and that of the graphic composition of non-geographic map elements. Good reference, the property by which the map reader associates labels with their referent geographic objects (points, lines, or regions) is the principal requirement of label placement. This leads to rules with a "fuzzy" character: the label (name) for a point feature must be "near" the point symbol for the same point feature, and should be unambiguously referable to that point symbol; linear feature labels (names) must be "near" the line symbols for the same linear features, and must be unambiguously related to the appropriate linear symbols; area symbols should cover approximately the same areas (regions) as do the areal features they represent.

Good reference is achieved primarily through the location of the name, but can be aided through the size, style, and color of the type chosen; choice of type is considered to be a symbolization issue (see above). Obviously, a large continuum of placements can achieve good reference for any particular object's label. However, labels take up a relatively large amount of map space, and one encounters crowding and space-competition even for small problems. As we will discuss below, labels and label placement may dominate selection for point features. Recently, Hinckley (1989) has reviewed some fundamental principles of cartographic typography.

Yoeli (1972) wrote an early point-feature labeling program, and Imhof (1972) provided many heuristics in his classic paper, "Positioning Names on Maps". Subsequent work has relied by and large on the heuristics presented in these works, and has concentrated on efficiency of algorithms. Kelley (1980) focused on efficient methods for determining potential name overlaps, and used grid- and quadtree-related techniques to subdivide the map area. Hirsch (1982) also focused on efficiency, but used vectors to represent and handle possible conflicts. Basoglu (1982) and Ahn and Freeman (Ahn and Freeman 1983; Ahn 1984; Freeman and Ahn 1984) were among the first to address the placement of point, line, and area feature labeling. Ahn's programs solved these in separate routines, written in FORTRAN. Pfefferkorn *et al.* (1985) re-implemented Ahn's methods in a different software environment.

Langran (Langran and Poiker 1986) provided a new direction to research by focusing attention on the interaction between selection of point features to be shown on the map (using Töpfer and Pillewizer's (1966) Radical Law) and the actual labeling of those features. For point symbols, the label almost invariably requires more space than the symbol itself, and so the labels will dominate the competition for map space and thus cartographic selection. Langran and Poiker (1986) report on Langran's program, which selected point features based on local importance, with a heuristic based on population potential. Mower (1986, 1988) used competition for map-space itself as the criterion for selection, and added spatially-dependent backtracking to the Imhof/Yoeli heuristics.

Cromley (1985, 1986) focused on linear programming optimization for point feature labeling. Zoraster (Zoraster 1986; Zoraster and Bayer (1987)) used similar optimization procedures. Van Roessel (1987) addressed the problem of finding appropriate places to put the labels for an areal feature (represented as a polygon), and included heuristics for choosing a label position if more than one candidate is found. Drinnan, Mattair and Luckley (1989) discuss an interactive editing system for labeling on nautical charts that uses an expert system to do initial placements. It contains rules expressed in a proprietary expert system language especially written for this purpose, and allows interaction with the user to refine the results.

Recently, four papers on automated cartographic name placement were presented at AUTO-CARTO 9. Collectively, these represent a departure from previous work, in that all address placement of point, line, *and* area feature labels. Ebinger and Goulette (1989) discussed a US Bureau of the Census system written in FORTRAN (F. Broome, personal communication, 1989). The system yields adequate results; Ebinger and Goulette suggest that improvements would best be accomplished interactively, rather than by refining the existing rules. (This provides another opportunity for implementing an amplified intelligence module within an expert system.)

A second system developed and described by Doerschler and Freeman (1989) includes a complex set of rules for selecting fonts, character sizes, etc., an aspect not explicitly discussed by other name-placement researchers. There are also many placement rules arranged in a sequence of desirability. After a placement is generated, its quality is measured; if the quality is not sufficiently high, the next placement rule is used, and so on until an acceptable placement is found or the name is deleted. The system is implemented as 700 FORTRAN modules. Johnson and Basoglu (1989) do not describe an existing system, but rather review new programming techniques, specifically neural networks, that may prove useful for name placement. Jones and Cook (1989) describe a rule-based system in which Prolog is used as the inference engine. The data base in their system is accessed by means of calls to FORTRAN subroutines. They plan to add an interface to allow interactive, natural language entry of new rules.

None of these authors has considered the fact that name labels must interact with all features on the map, and not just with other labels and features of the same kind. Also, except for the system written by Jones and Cook (1989), none have separated the data base, rule base, and inference engine, and their systems do not have capabilities for explanation of particular solutions. A major part of the research agenda for developing a cartographic expert system is to extract the rules from the programs and articles discussed above, to put them into a standard form, and to enter them into an expert system shell for evaluation and eventually integration with other cartographic expert system modules.

Another part of the research agenda involves the evaluation of the rules, and the maps they produce, and the search for additional rules and heuristics. Evaluation should be based on perceptual testing and performance evaluation. Empirical examinations of existing maps also will contribute to this second subagenda. Thus far, there has been very little testing of user performance on name-related tasks; an exception is Eastman's (1985) study of the interaction between map-based learning and name placement. His conclusion was that people remembered town locations with the lowest spatial error if names were centered immediately above or below the point symbol, a conclusion that conflicts with the Imhof/ Yoeli heuristics on preferred placement. However, it is probable that most maps are intended to be used directly through perception, and map memory may be less relevant.

There is a critical need for an evaluation of user performance on maps generated under different placement heuristics. But first, a formal analysis of the tasks, the things that map readers use the names for, must be conducted. Empirical investigations of name-placement practice also have been rare. We know of only two such studies, both recent, unpublished Master's research projects at our University (Wu 1989; Christen 1989). Further research is clearly needed to validate map design guidelines empirically.

5. *Visual hierarchy and contrast.* After the map locations of geographic objects and of annotations have been determined, the map must be tested for overall graphic quality (balance, figure-ground, symmetry, and contrast). If deficiencies are detected, problems should be alleviated without violating any of the cartographic principles identified above. In the course of this operation, the non-geographic components of the map, including titles, legends, north-arrows, scale-bars, etc., and the map border itself, are placed in order to achieve appropriate balance and appearance. Colors, values, and visual texture might also be adjusted at this stage.

If adequate visual contrast cannot be achieved by these adjustments, other aspects of layout, composition, and annotation would have to be modified. In principle, the process might occasionally have to backtrack all the way to the generalization module. Little is understood about our comprehension of the visual hierarchy, although the work of Arnheim (1971), Bertin (1983), and Dent (1985) provides many guidelines for principles of manual design.

Progress to Date

Not all phases of map design require an expert systems approach. Figure 7.3 summarizes our estimate of the potential role of an expert systems approach for each of the major components of the map design process discussed in this chapter. For those with medium to high potential, we also estimate the

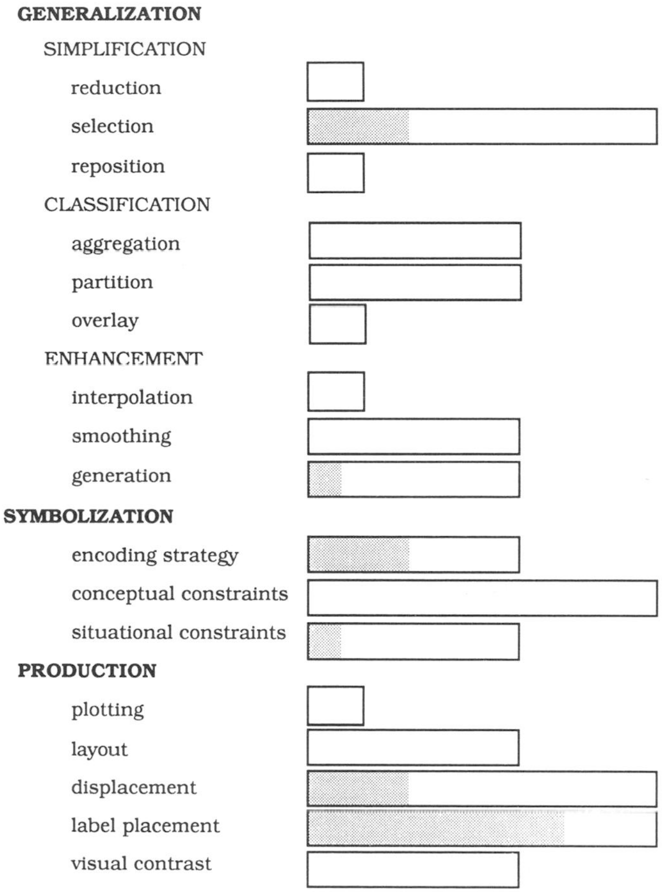

GENERALIZATION
SIMPLIFICATION
reduction
selection
reposition
CLASSIFICATION
aggregation
partition
overlay
ENHANCEMENT
interpolation
smoothing
generation
SYMBOLIZATION
encoding strategy
conceptual constraints
situational constraints
PRODUCTION
plotting
layout
displacement
label placement
visual contrast

FIG. 7.3. Role of expert systems in map design (revised from Mark and Buttenfield 1988).

degree of progress to date. In the figure, the length of the bar estimates the applicability (low, medium, or high) of an expert systems approach to various components of map design. Shaded areas indicate the proportion (none, small, medium, or large) of that potential achieved to date. Several items in this framework are notable in terms of research progress towards an expert system approach.

The first area of research progress lies in the area of map generalization. Straightforward algorithms are effective for line reduction, repositioning, interpolation, and plotting. Geometrical (pattern recognition) and geographical (substantive) decision-making play key roles in cartographic selection to select and modify tolerance values during simplification. Classification including aggregation and partitioning requires knowledge, but may benefit from an amplified intelligence approach (user interaction), and need not be done "on the fly". Straightforward algorithms can produce adequate enhancement to reconstruct generic cartographic lines and surfaces, but could be improved through an expert systems approach, to simulate detail and texture for specific types of geographic phenomena.

A second fertile area lies in the realm of symbology. The encoding of symbol choice and type choice based on semiotic considerations is a fertile yet untaped area for expert systems. Heuristics for the selection among various alternatives for representing thematic data (such as choroplethic or isoplethic shading, cartograms, etc.) also should be developed and tested. Choice of colors and especially color sequences are based on important conceptual constraints that may prove readily adapted as expert system rules. Guidelines for psychophysical compensation have not been widely explored in the cartographic literature; but once determined they will also prove straightforward to integrate. Situational constraints are most often based on map purpose and the map reading audience; we believe the potential for expert system development in this area is quite low.

In map production one discovers the majority of research to date applying expert systems approaches. Displacement is an important research area, but has been addressed only by Nickerson and Freeman (1986). Label placement is the principle area of research to date; much has been achieved, but the problems of embedding features with labels into the map fabric remains a real and pressing issue. The question of visual balance and contrast has not been addressed in the current literature. We know of no work on determining measures of the overall visual appearance of a map, or of placing non-geographic elements according to principles of graphic design. As so many design principles are taught by example, this area will prove highly challenging to automate in an expert system shell.

Summary and Further Progress

The actual development of a full cartographic expert system will require the work of cartographers, programmers, and knowledge engineers, over a period of several to many years. First, a small team must develop a very clear design for the system. The choice of an effective data structure is especially important. The design must be highly modular, so that elements may be developed as independent modules, to be combined into the evolving cartographic expert system. During the development phase, some modules might be implemented using amplified intelligence methods, while other design tasks for which rules may be readily formalized could be implemented in a full expert systems approach. For those parts of the problem that have been effectively addressed in research reported in the literature, programs must be converted to the selected programming environment, data structures, and system design adopted. Point feature labelling, and selection of tolerance values in simplification of cartographic lines might be two good problems to address initially. It is important to evaluate the final products of any automated system by means of map user response, and this area of research will require careful consideration throughout the next decade.

Acknowledgements

This paper represents part of Research Initiative No. 8, "Cartographic Expert Systems", of the National Center for Geographic Information and Analysis, supported by a grant from the National Science Foundation (SES-88-10917); support by NSF is gratefully acknowledged. Some of the ideas expressed here were outlined briefly in a conference presentation (Mark and Buttenfield 1988). The authors can be reached via the BITNET electronic mail system: Mark's address is geodmm@ubvms, and Buttenfield's is geobabs@ubvms.

References

Ahn, J. A. (1984) *Automatic Name Placement Systems*. Image Processing Laboratory Report IPL-TR-063, Rensselaer Polytechnic Institute, Troy, New York, 83 pp.
Ahn, J. A. and H. Freeman (1983) "A program for automated name placement", *Proceedings Sixth International Symposium on Computer-Assisted Cartography (Auto-Carto 6)*, Ottawa, pp. 444–453.
Arnheim, R. (1971) *Visual Thinking*. University of California Press, Berkeley.
Basoglu, U. (1982) "A new approach to automated name placement", *Proceedings, Fifth International Symposium on Computer-Assisted Cartography (Auto-Carto 5)*, pp. 103–112.
Beard, C., F. Broome and A. Martinez (1987) "Automated map inset determination", *Proceedings, Eighth International Symposium on Computer-Assisted Cartography (Auto-Carto 8)*, Baltimore, pp. 466–470.
Beard, M. K. (1987b) "Data descriptions for automated generalization", *Proceedings, International Geographic Information Systems (IGIS) Symposium: The Research Agenda*, Vol. 2, pp. II-3 to II-10.
Beard, M. K. (1987a) *Multiple Maps from a Detailed Data Base: A Scheme for Automated Generalization*. Unpublished Ph.D. Dissertation, University of Wisconsin, Madison.
Bertin, J. (1983) *Semiology of Graphics*. University of Wisconsin Press, Madison.
Brachman, R. J., S. Amarel, D. Engelman, R. S. Engelmore, E. A. Feigenbaum, and D. E.

Wolkins (1983) "What are expert systems?" In Hayes-Roth, F., D. A. Waterman and D. B. Lenat (eds.), *Building Expert Systems*. Addison-Wesley, Reading, Massachusetts, pp. 31–57.

Brassel, K. E. and R. Weibel (1987) "Map generalization". In Anderson, K. E. and A. V. Douglas (eds) *Report on International Research and Development in Advanced Cartographic Technology*, 1984–1987, International Cartographic Association, pp. 120–137.

Brownston, L., R. Farrell, E. Kant, and N. Martin (1985) *Programming Expert Systems in OPS5: An Introduction to Rule-Based Programming*. Addison-Wesley, Reading, Massachusetts.

Buttenfield, B. P. (1984) *Line Structure in Graphic and Geographic Space*, unpublished Ph.D. dissertation, University of Washington, Seattle.

Buttenfield, B. P. (1987) "Automating the identification of cartographic lines", *The American Cartographer*, Vol. 14, No. 1, pp. 7–20.

Buttenfield, B. P. (1989) "Scale-dependence and self-similarity in cartographic lines", *Cartographica*, Vol. 26, No. 1, pp. 79–100.

Child, J. (1984) *A Semiotic Approach to Cartographic Structure and Map Meaning*, unpublished Ph.D. dissertation, University of Washington, Seattle.

Christen, L. (1989) *The Influence of Position Rankings on Point Name Placement for Manually Produced Road Maps*, unpublished master's thesis, State University of New York at Buffalo, Buffalo, New York.

Cromley, R. G. (1985) "An LP relaxation procedure for annotating point features using interactive graphics", *Proceedings, Seventh International Symposium on Computer-Assisted Cartography (Auto-Carto 7)*, pp. 127–132.

Cromley, R. G. (1986) "A spatial allocation analysis of the point annotation problem", *Proceedings, Second International Symposium on Spatial Data Handling*, Seattle, Washington, pp. 38–49.

Cuff, D. J. and M. T. Mattson (1982) *Thematic Maps: Design and Production*, New York, Methuen.

Dent, B. D. (1985) *Principles of Thematic Map Design*, Addison-Wesley, Reading, Mass.

Doerschler, J. S. and H. Freeman (1989) "An expert system for dense-map placement", *Proceedings, Ninth International Symposium on Computer-Assisted Cartography (Auto-Carto 9)*, Baltimore, pp. 215–224.

Drinnan, C. H., C. G. Mattair, Jr. and S. E. Luckley (1989) "An interactive editing system for nomenclature placement", *Technical Papers, 1989 ASPRS/ACSM Annual Convention*, Vol. 5, pp. 221–230, Baltimore.

Dutton, G. H. (1981) "Fractal enhancement of cartographic line detail", *The American Cartographer*, Vol. 8, pp. 23–40.

Eastman, J. R. (1985) "Name placement and positional recall of map information", *Technical Papers, 45th Annual Meeting ACSM*, pp. 474–482.

Ebinger, L. R. and A. M. Goulette (1989) "Automated names placement in a non-interactive environment", *Proceedings, Ninth International Symposium on Computer-Assisted Cartography (Auto-Carto 9)*, Baltimore, pp. 205–214.

Fisher, P. and W. Mackaness (1987) "Are cartographic expert systems possible?" *Proceedings, Eighth International Symposium on Computer-Assisted Cartography (Auto-Carto 8)*, Baltimore, pp. 530–534.

Flannery, J. J. (1971) "The relative effectiveness of some common graduated point symbols in the presentation of quantitative data", *The Canadian Cartographer*, Vol. 8, pp. 96–109.

Frank, A. U., D. L. Hudson and V. B. Robinson (1987) "Artificial intelligence tools for GIS", *Proceedings, International Geographic Information Systems (IGIS) Symposium: The Research Agenda*, Vol. 2, pp. II-257 to II-271.

Freeman, H. and J. Ahn (1984) "AUTONAP – An expert system for automatic name placement", *Proceedings, International Symposium on Spatial Data Handling*, Zurich, 1984, pp. 544–569.

Ganter, J. H. (1988) "Interactive graphics: Linking the human to the model", *Proceedings, GIS/LIS '88*, Vol. 1, pp. 230–239.

Gilmartin, P. P. (1978) "Evaluation of thematic maps using the semantic differential test", *The American Cartographer*, Vol. 5(2), pp. 133–139.

Heyn, B. N. (1981) "An evaluation of map color schemes for use on CRTs". Unpublished

Master's thesis, Department of Geography, University of South Carolina, Columbia, South Carolina.

Hinckley, T. K. (1989) "The canons of cartographic typography", *Technical Papers, 1989 ASPRS/ACSM Annual Convention*, Vol. 5, pp. 212–220, Baltimore.

Hirsch, S. A. (1982) "An algorithm for automatic name placement around point data", *The American Cartographer*, Vol. 9, pp. 5–17.

Imhof, E. (1972) "Positioning names on maps", *The American Cartographer*, Vol. 2, pp. 128–144.

Jenks, G. F. (1963) "Generalization in statistical mapping", *Annals, Association of American Geographers*, Vol. 53, pp. 15–26.

Johnson, D. S. and U. Basoglu (1989) "The use of artificial intelligence in the automated placement of cartographic names", *Proceedings, Ninth International Symposium on Computer-Assisted Cartography (Auto-Carto 9)*, Baltimore, pp. 225–230.

Jones, C. B. and A. D. Cook (1989) "Rule-based cartographic name placement with Prolog", *Proceedings, Ninth International Symposium on Computer-Assisted Cartography (Auto-Carto 9)*, Baltimore, pp. 231–240.

Kelley, P. C. (1980) *Automated Positioning of Feature Names on Maps*, unpublished Master's thesis, State University of New York at Buffalo, Buffalo, New York.

Kimerling, A. J. (1975) "A cartographic study of equal value gray scales for use with screened gray areas", *The American Cartographer*, Vol. 2(2), pp. 119–127.

Langran, G. E. and T. K. Poiker (1986) "Integration of name selection and name placement", *Proceedings, Second International Symposium on Spatial Data Handling*, Seattle, Washington, pp. 50–64.

Leonard, J. J. and B. P. Buttenfield (1989) "An equal value gray scale for laser printer mapping", *The American Cartographer*, Vol. 16(2), pp. 97–107.

Mackaness, W. A. and P. F. Fisher (1987) "Automatic recognition and resolution of spatial conflicts in cartographic symbolization", *Proceedings, Eighth International Symposium on Computer-Assisted Cartography (Auto-Carto 8)*, Baltimore, pp. 709–718.

Marino, J. S. (1979) "Identification of characteristic points along naturally occurring lines: an empirical study", *Canadian Cartographer*, Vol. 16(1), pp. 70–80.

Mark, D. M. (1989) "Conceptual basis for geographic line generalization", *Proceedings, Ninth International Symposium on Computer-Assisted Cartography (Auto-Carto 9)*, Baltimore, pp. 68–77.

Mark, D. M. and B. P. Buttenfield (1988) "Design criteria for a cartographic expert system", *Proceedings, 8th International Workshop on Expert Systems and Their Applications*, Avignon, France, June, Vol. 2, pp. 413–425.

Martinez, A. A. (1989) "Automated insetting: An expert component embedded in the Census Bureau's map production system", *Proceedings, Ninth International Symposium on Computer-Assisted Cartography (Auto-Carto 9)*, Baltimore, pp. 181–190.

McKay, J. R. (1949) "Dotting the dot map", *Surveying and Mapping*, Vol. 9, pp. 3–10.

McMaster, R. B. (1987) "Automated line generalization", *Cartographica*, Vol. 24, pp. 74–111.

McMaster, R. B. and K. S. Shea (1988) "Cartographic generalization in a digital environment: a framework for implementation in a geographic information system", *Proceedings, GIS/LIS '88*, Vol. 1, pp. 240–249.

Monmonier, M. S. (1988) "Interpolated line generalization: overview, examples, and preliminary evaluation", *Proceedings, GIS/LIS '88*, Vol. 1, pp. 256–265.

Mower, J. E. (1986) "Name placement of point features through constraint propagation", *Proceedings, Second International Symposium on Spatial Data Handling*, Seattle, Washington, pp. 65–73.

Mower, J. E. (1988) *The Selection, Implementation, and Evaluation of Heuristics for Automated Cartographic Name Placement*, unpublished Ph.D. dissertation, State University of New York at Buffalo, Buffalo, New York.

Muller, J. C., R. D. Johnson and L. R. Vanzella (1986) "A knowledge-based approach for developing cartographic expertise", *Proceedings, Second International Symposium on Spatial Data Handling*, Seattle, Washington, pp. 557–571.

Nickerson, B. G. and H. Freeman (1986) "Development of a rule-based approach for automatic map generalization", *Proceedings, Second International Symposium on Spatial Data Handling*, Seattle, Washington, pp. 537–556.

Nyerges, T. L. and Jankowski, P. (1989) "A knowledge base for map projection selection", *The American Cartographer*, Vol. 16, pp. 29–38.

Olson, J. M. (1987) "Color and the computer in cartography", Chapter 11, pp. 205–221 in *Color and The Computer*, H. John Durrett (ed). Academic Press, Boston.

Pannekoek, A. J. (1962) "Generalization of coastlines and contours", *International Yearbook of Cartography*, Vol. 2, pp. 55–74.

Patton, J. C. and P. V. Crawford (1966) "The perception of hypsometric colors", *The Cartographic Journal*, Vol. 13, pp. 115–127.

Petchenik, B. B. (1974) "A verbal approach to characterizing the look of maps", *The American Cartographer*, Vol. 1, pp. 63–71.

Peuquet, D. (1984) "Data structures for a knowledge based geographic information system", *Proceedings, International Symposium on Spatial Data Handling*, Zurich, 1984, pp. 372–391.

Pfefferkorn, C., D. Burr, D. Harrison, B. Heckman, C. Oresky and J. Rothermel (1985) "ACES: a cartographic expert system", *Proceedings, Seventh International Symposium on Computer-Assisted Cartography (Auto-Carto 7)*, pp. 399–407.

Robinson, G. and M. Jackson (1985) "Expert systems in map design", *Proceedings, Seventh International Symposium on Computer-Assisted Cartography (Auto-Carto 7)*, pp. 430–439.

Robinson, A. H., R. D. Sale, J. L. Morrison and P. C. Muehrcke (1984) *Elements of Cartography* [5th edition], Wiley, New York. Schlictmann, H. (1985) "Characteristic traits of the semiotic system 'map symbolism'", *The Cartographic Journal*, Vol. 22, pp. 22–30.

Shea, K. S. and R. B. McMaster (1989) "Cartographic generalization in a digital environment: when and how to generalize", *Proceedings, Ninth International Symposium on Computer-Assisted Cartography (Auto-Carto 9)*, Baltimore, pp. 56–67.

Shortridge, B. G. (1979) "Map reader discrimination of lettering size", *The American Cartographer*, Vol. 6(1), pp. 13–20.

Smith, R. M. (1987) "Influence of texture on perception of gray tone map symbols", *The American Cartographer*, Vol. 14(10), pp. 43–47.

Smith, T. R. (1984) "Knowledge based control of search and learning in a large-scale geographic information system", *Proceedings, International Symposium on Spatial Data Handling*, Zurich, pp. 498–519.

Smith, T. R. and M. I. Pazner (1984) "A knowledge-based system for answering queries concerning geographic objects", *Proceedings, Pecora 9* (IEEE), pp. 286–289.

Töpfer, F. and W. Pillewizer (1966) "The principles of selection", *The Cartographic Journal*, Vol. 3, pp. 10–16.

Tufte, E. R. (1983) *The Visual Display of Quantitative Information*, Graphics Press, Cheshire, Connecticut.

Van Roessel, J. W. (1987) "An algorithm for locating candidate boxes within a polygon", *Proceedings, Eighth International Symposium on Computer-Assisted Cartography (Auto-Carto 8)*, Baltimore, pp. 689–700.

Weibel, R. and B. P. Buttenfield (1988) "Map design for geographic information systems", *Proceedings, GIS/LIS '88*, Vol. 1, pp. 350–359.

White, E. R. (1985) "Assessment of line generalization algorithms using characteristic points", *The American Cartographer*, Vol. 12(1), pp. 17–27.

Williams, R. L. (1960) "Map symbols: The curve of the gray spectrum—an answer", *Annals, Association of American Geographers*, Vol. 50, pp. 487–491.

Wu, C. V. (1989) *Verification of Rules for Name Placement on Maps*, unpublished master's thesis, State University of New York at Buffalo, Buffalo, New York.

Yoeli, P. (1972) "The logic of automated map lettering", *The Cartographic Journal*, Vol. 9, pp. 99–108.

Zhang, W., H. Li and X. Zhang (1988) "MAPGEN: An expert system for automatic map generalization", *Proceedings of the 13th International Cartographic Conference*, Morelia, Mexico, October 12–21, Vol. 4, pp. 151–157.

Zoraster, S. (1986) "Integer programming applied to the map label placement problem", *Cartographica*, Vol. 23, pp. 16–27.

Zoraster, S. and S. Bayer (1987) "Practical experience with a map label placement program", *Proceedings, Eighth International Symposium on Computer-Assisted Cartography (Auto-Carto 8)*, Baltimore, pp. 701–708.

CHAPTER 8

Digital Geographic Interchange Standards

TIMOTHY V. EVANGELATOS

Canadian Hydrographic Service
Fisheries and Oceans
Ottawa, Canada

Introduction

The lack of (useful) digital geographic interchange standards is seen as a significant impediment to the growth of geographic information systems (GIS) among general users since the incorporation of data from others can be an expensive, time consuming and very frustrating process. Substantial effort is now being made in many countries to overcome these difficulties through the development of national standards. A number of proposals have appeared but there is no indication that any one of these proposals will be widely accepted in the international community.

This chapter discusses some of the issues retarding the development and acceptance of such standards, and the various areas where standardization is required. It then reviews the current national and international efforts to develop digital geographic and cartographic interchange standards.

Major difficulties in developing useful standards have been the lack of a standard terminology which would aid the geomatics community in understanding and communicating what the problems really are, and the lack of a mature science of geomatics which could provide a sound basis for designing good standards. Fortunately, both of these problems are slowly disappearing.

Standards Issues

This section discusses several of the issues which have prevented the development of useful and generally acceptable geographic exchange standards. Not surprisingly the task of data exchange has been grossly oversimplified. In the 1970s it was adequate to transfer coordinates, a task which appeared relatively easy, but as the data became more sophisticated the "digital map" grew from strings of coordinates (spaghetti) to shared line segments (chain node) without gaps, overshoots or slivers (structured data) to a complete topological description of all nodes, lines and polygons. From feature-coded features, flexible attributional schemes were added which led to the incorporation of relational techniques and finally to an object-oriented approach. Initially data exchange was the preserve of the national mapping agencies using bulk magnetic tape, but for the 1990s the distribution of data to and between users (and value adders) will become much more important and probably dominate the exchange of digital geographic information. During the past few years new media such as telecommunications, CD-ROMS and Laser Video disks have also been introduced and have begun to replace magnetic tapes for the dissemination and exchange of geo-information, particularly for microcomputer-based systems (see chapters by Lee and Coll in this volume). In the future, dissemination of bulk data will likely be via CD-ROMs and update information through telecommunications networks.

The potential benefit of creating standards, whether through a consensus process or by regulation, can be enormous. Although the recognition of this fact has stimulated the demand for better standards, generally useful and acceptable standards are still not available. The reasons for this apparent paradox are as follows:

Evolution and Change

The ongoing evolution of the science and technology underlying spatial science continues to promise greater things; similarly the applications using the technology are becoming broader and the databases needed to serve the applications based upon it are also evolving. It is also perceived that more integration with other engineering disciplines will soon be required since they are making more and more use of GIS technology; for example the use of CAD/CAM systems for road design.

Spatial Science

The basic foundations of spatial science are still incomplete and there is no fully adequate theoretical basis upon which to engineer good, complete standards. This also leads to an incomplete understanding as to how best to apply and use the technology.

Technology

Even though the hardware and software underlying geomatics are changing fairly rapidly (see chapters by Lee, Keller and Waters, Peuquet, and

Buttenfield and Mark in this volume), the GIS systems now available are not fully developed. This has forced some users to employ innovative but potentially non-standard methods and designs for solving their problems. This creates difficulties for the user in adopting standards.

User Commitment

Lack of commitment among users to agree on standards because of: Dissatisfaction with current proposals; belief that something better is around the corner, and it is; vested interests in current defacto standards; and the high cost of changing to a new standard. Even when good standards are available, users often cannot afford to change.

Vendor Motivation

Lack of motivation among vendors and manufacturers to cooperate in the development of industry standards because of: High cost, high risk and long-term return on investment; fear of aiding competitors through open systems and losing the advantageous position of having customers locked into one system; business accruing from the sale of conversion software and services; and lack of economic and other pressures from users.

Future Visions

Differing visions of the future makes it difficult for users to agree on specifications of proposed standards. By the time consensus is reached, new players are pushing new visions that cannot be ignored.

Knowledge

Expertise of standards developers is often limited or narrowly focused. For example, the tremendous effort under way by communications engineers for the development of Open Systems Interconnection (OSI), which will have an enormous impact upon computer-to-computer exchanges of the future, has until recently been largely ignored by developers of geographic exchange standards.

Content

What should be exchanged? Fundamental issues such as whether topological relations should be included in the transfer have not been adequately resolved. The specification of the underlying model or models (e.g. relational, network, object-oriented) is still being debated, and the best way of classifying and describing features is far from being resolved.

Do Everything

Solutions that attempt to include everything have not proven to be practical. Similar solutions that are very flexible are not "standards" unless additional layers of constraint are imposed. The paradox is: If it is easy for the sendor to encode data into the standard then it is difficult for the receiver to decode and use the data. Conversely, to make it easy for the receiver to use the data it becomes harder for the sender to encode the data. An appropriate balance has not yet been found.

Not Invented Here

Many agencies or groups having their own solution or view of the exchange requirement makes it difficult to reach a consensus. These aspects are further complicated when careers and budgets are also tied to specific standards projects.

Cost

The development, specification, testing and support involved with creating standards is very expensive. Generally only governments have the resources to carry out the task.

Isolated Efforts

Most national organizations are working in some degree of isolation. Furthermore, none of their solutions have received any commercial push that might crystallize their use and make them into a widely used standard.

Thus there are many reasons why broadly acceptable geomatic standards have not appeared. There is no strong evidence that any of the solutions now under development (and reviewed later) will become generally acceptable. Many users are continuing to "make do" with CAD/CAM standards such as ISIF (Intergraph Standard Interchange Format), which has been in use for well over a decade, and DXF, a newer CAD/CAM standard that has become very popular with microcomputer users. Some CAD, graphic and office document standards (Rainio 1989; Salgé 1989) are being extended to cover mapping, and it is quite possible that some of them will make significant inroads into the geomatics area because of the much larger community of users in other disciplines who will want to gain access to spatial data.

Areas of Standardization

Early attempts to develop exchange standards tended to solve the exchange problem in one all-encompassing solution. Most of these solutions have proven to be unsatisfactory for exchanging anything more than raw unstructured data from digitizing sources. Current approaches involve partitioning the problem into separate disjointed areas: For example the separation of the content from the packaging; the coordinate data from the feature coding; and symbolization from the attribution data. This is a very important step since it allows each component to be independently developed and optimized. The various areas of standardization are as follows (Evangelatos *et al.* 1990).

Feature Definition and Classification

The definition of all the features or objects (and synonyms) and their attributes. Although it takes a great deal of effort, most standards bodies attempt to classify features in a hierarchical structure. This approach has met with limited success since most users have differing views as to how the features should be grouped. For the exchange process there appears to be a trend away from structuring the features in any but very simple classification schemes.

Cataloguing of Datasets

The reference system(s) to be used to build the libraries and on-line directories for the datasets. This facilitates access to the datasets and awareness of what is available in these datasets. It should build upon existing cataloguing and retrieval mechanisms and can therefore use tools already in place in the library world.

Data Models

The representation of spatial features and the interrelationships among the features that are to be stored. These can become fairly complicated. Standard data models for databases and for data exchange would aid the designers of those systems. Rules for the digital representation of more complex features that consist of more than a single point, line or area are also important if the features are to be properly recreated without loss of information.

Data Encoding

This refers to how the data is packaged for exchange, whether for a direct access device like a floppy disk or CD-ROM or for a sequential telecommunications link.

The following list indicates the different types of data that are being considered:

— Vector Data (spaghetti, structured data, planar topological, objects).
— Raster Data (Remote Sensed Data and Scanned Map and Chart images). In the past, data types such as raster and vector have been

treated in separate standards or in separate modules of the same standard, but the trend is for the integration of these two data types (see chapter by Peuquet in this volume). Future standards must allow for this.

— Matrix/Gridded Data (e.g. Digital Elevation Models).[1]
— Video Map and Chart Images.
— CAD (growing number of applications in engineering needing access to geographic information).

Two complementary information processing standards from the International Standards Organization (ISO) have gained significant support for the encoding of geomatics data (O'Brien 1988). These are ISO 8211 for physical media (ISO 1985a, Specification for Data Descriptive File for Information Interchange), and ISO 8824/5 (ISO 1985b, OSI Specification of Abstract Syntax Notation One (ASN 1)).

Furthermore, the ISO has devised a concept of Open Systems Interconnection (OSI). This is a framework within which protocols for communication between dissimilar computer systems can be devised (Voelker 1986). It divides the computer to computer exchange process into seven separate compartments or layers, each dealing with a specific aspect of the complex communication problem. This model and the standards that have been developed under it are getting world wide acceptance, and therefore it is recommended that it also be used as a basis for geographic data exchange formats. Unfortunately ISO 8211, which appears to be gaining popularity in the geographic world is not, in its current form, a part of the Open Systems Interconnection model.

Relationships Between Features (Objects)

The relationships that need to be carried out are becoming more complicated. Some examples are:

— Spatial relationships (i.e. topological).
— Non-spatial relationships. In systems being considered for the future there appears to be a requirement to link other information which may not have a explicit spatial position. An example is the prototype hypermedia electronic navigation chart developed by the Canadian Hydrographic Service which incorporates the chart data with sailing directions, photographs, video, sound and other information of interest to the navigator (Benninger 1988).
— Both spatial and non-spatial. For example, there is a trend to allow for the creation of higher order objects from other (atomic) objects in order to provide a better model of reality.

---

[1] Often treated as a form of raster structure.

Geographic Referencing

This obviously applies to the geographic coordinates such as latitude and longitude and the spheroids used to generate these coordinates. It could also apply to many other referencing schemes such as the Zip Code system of postal agencies or street names and address systems used for referencing properties. Good standards for geographic referencing exist, but many users are slow to adopt them because of the high cost of conversion from their existing framework. For marine navigation a need also exists for international standardization of the vertical or tidal datum.

Data Quality

This is the "fitness for use" of the data and includes age, completeness (of a data set), accuracy, consistency and precision. Standard methods to measure and record the quality of the digital data exist for some data such as geodetic measurements, and for which the accuracy specification (e.g. error ellipse) may be easy to specify, but for generalized, selected map data it will be difficult to develop useful comprehensive standards.

Symbology

Standards for the drawing of spatial data may not be too important, but when on-line terminals are in general use for interrogating geographic databases, it will be useful to have standards for symbolizing and drawing the base maps for frequently used applications. An international standard for symbolization is essential for the electronic navigational chart.

Terminology

Even though the situation has improved, much confusion still exists with the use of many terms adopted from mathematics, computer science, surveying, mapping, communications and other fields.

The exchange of geomatic data is seen as a pressing problem and there tends to be a general belief that a good exchange format will fully resolve this problem. This may be a myth because in many cases the receiver must understand how the user has described and digitally modelled each feature if the received data is to be used correctly. This implies the need for greater standardization of the data bases themselves.

Overview of Standards Activities

As the previous section indicated, standards are needed over a broad range of areas. Many nations are now devoting considerable effort to develop national standards and there have been at least two attempts to create international standards.[2] This section gives a brief overview of several of these national standards activities. It is partly based upon the presentations given at meeting of the International Cartographic Association (ICA) Working Group on Digital Cartographic Standards, held on August 17 and 21, 1989 in Budapest. It also describes the efforts of two groups to create international standards, and concludes by attempting to predict some possible directions that developments may take.

National Activities

Australia

A national format for the interchange of feature-coded digital map data (via magnetic tape) was issued by the Standards Association of Australia (SAA) in 1981. It was based upon the work of a National Mapping Council Working Party that had been created for that purpose and reflects the state-of-the-art computer-assisted map production of the late 1970s. The format was revised in 1984 by adding more feature codes and improving the scope of the record types, with the final version published in 1988. The format is widely used by government agencies in Australia but primarily for the transfer of digitized map data from private sector firms working under government contracts. Like formats of that era its capabilities for identifying features is inadequate and it is not suitable for database and GIS applications. Therefore, in 1987 the SAA created a new committee, the IT/4, to consider the more general requirements of Geographic Information Systems. It consists of three subcommittees: IT/4/2, Geographic Data Exchange Formats; IT/4/3, Bibliographic Elements on Maps; and IT/4/4, Entity and Attribute Definitions, which has formed working groups for Land Use Data; Geological Data; Natural Resources Data; Cadastral Data; Street Addressing; and Utilities.

Plans are being made for a new standard which is to incorporate ISO 8211 and it will probably be an adaptation of the American Spatial Data Transfer Specification (SDTS) for Australian requirements. Implementation is planned for 1990, assuming the U.S. effort is completed (Lindsay 1989).

Canada

Under the Canadian Council on Surveying and Mapping (CCSM), an organization of federal and provincial surveyors and mappers, a technical committee has developed and published the following suite of standards to cover topographic mapping: Quality Evaluation of Digital Topographic

---

[2] This ignores existing international standards in specialized areas such as oceanography and remote sensing.

Data (CCSM 1984a); Classification and Coding of Features (CCSM 1984b); Standard EDP File Exchange Format for Digital Topographic Mapping (CCSM 1988); and A Digital Topographic Information Model (CCSM 1986).

The specifications were designed for the distribution of topographic map data and is used by the Department of Energy, Mines and Resources (EMR) for the distribution of digital map data. EMR is also responsible for maintaining the specifications and aiding users in implementing the specifications. Transfer is by magnetic tape and covers vector data only. The format was recently renamed as the Canadian Council on Geomatics Exchange Format (CGOGEF). The principal applications are for the distribution of topographic data from EMR.

Canada has a national standards system rather than a single national standards organization, as is the case in most countries. At the core of this standards system is the Standards Council of Canada which accredits and coordinates the efforts of other standards-writing organizations, such as the Canadian General Standards Board (CGSB). In 1989 a Committee on Geomatics for the development of national standards was established under the auspices of the CGSB with an overall objective to develop standards to promote the sharing of geomatics data. Four working groups were established for: Data Transfer/Interchange Formats, (WG1); Data Models for the Transfer Format, (WG2); Classification of Features, (WG3); Cataloguing of Data Sets; and Data Dictionary/Directory, (WG4). The thrust is to take a highly modular approach to solving the exchange problem with a clear separation between the contents (spatial information) and the carrier (packaging or encoding).

In 1985 a project was initiated to take a different approach to the encoding of spatial data. It produced MACDIF, for Map and Chart Data Interchange Format Version 2.0 (CHS 1988), which was developed jointly by the Ontario Ministry of Natural Resources (OMNR) and the federal government. It was the first such geographic exchange format to be built on the Open Systems Interconnection (OSI) concept of the International Standards Organization (ISO). In 1988, due to lack of resources and differing short term goals, OMNR began work on an enhancement of the MACDIF specification to include thematic attributes; a new version is in preparation (OMNR 1989). The work of the federal government has continued separately and MACDIF has been revised to Version 2.3 (CHS 1990) to include an updating mechanism.

Federal Republic of Germany

Completion of the integration of the European Economic Community in 1992 is expected to lead to an increased sharing of digital geographic data. FRG sees an urgent need for the harmonization of standards for Western

Europe. National and international communications networks will be created for the exchange of information not only for mapping but also for planning, environmental protection, defence, policing, automobile traffic and many other applications. At present no suitable standard exists to meet all the perceived requirements, but adaptation of the American SDTS specification is a possibility (Brüggemann 1989).

Standards are the responsibility of the Working Committee of the Surveying Administration of the Länder of FRG, which created a German Standard Data Format for cartographic data exchange in 1972. This was followed by another format (EDBS, Unified Database Interface) in 1982 for cadastral mapping. To deal with the evolving need for information for land navigation systems, a new standard has been designed. Called the Geographic Data File (GDF), it is based upon the British NTF but adapted to European needs. Version 1.0 was published in 1988 and has been going through an evaluation phase during the past year (Claussen *et al.* 1989).

Finland

The establishment of geomatic standards in Finland appears to be fairly advanced. In 1985 the Ministry of Agriculture and Forestry established the LIS-Project (Vahala 1986) to develop standards for modelling and classifying and exchanging land information data. The goal was to develop a system for joint use of geo-information based upon decentralized GIS systems (Rainio 1989). Approximately thirty-one agencies have been involved in this work and they have produced a transfer standard and a spatial data dictionary.

The exchange standard is based upon EDIFACT (ISO 9735), which is part of the Electronic Document Interchange (EDI) which, in turn, forms part of an even more general model for Office Documentation Architecture (ODA) that is being created under the open system interconnection concept of ISO.

The spatial data dictionary gives detailed information about what data is available in the various decentralized organizations. It describes the owners, contents, areal coverage as well as the feature types, the classification used (which differs from agency to agency), and the description of the geometric objects.

Conversion software is required between each GIS and the EDIFACT format. A tabular conversion is done when the user and receiver have different codings for the same feature, and coordinate transformations are done if different coordinate systems are used. The data dictionary plays a very important role in the exchange since the user must know who provides the data so he can perform the appropriate conversions when he gets the data. This approach does not support "blind interchange", but it is an important step and allows data suppliers to define standard products.

France

The National Council for Geographic Information (CNIG) established a Working Group of the Permanent Commission for Geographic Research (CPRG) in May, 1988 to investigate the definition of a national exchange format (Salgé 1989). This group was charged with defining the logical and physical structures for exchange; the feature coding system; terminology; and the standard itself.

The standard is to be based upon other national mapping standards or on international standards that are actually supported by systems manufacturers. Some of the possibilities are an ISO 8211 version of NTF, the British format; EDIFACT, an OSI based standards for commercial and shipping applications but which would require the definition of a geographic layer (also used by Finland); and a CAD/CAM standard STEP, which would also require the definition of a geographic layer. The group is aiming at developing a European Standard.

Japan

The development of a single general purpose exchange standard has not been attempted. Instead, several formats have been proposed based upon application and map scale. Features are stored in one of twenty-five standard layers. Exchange is currently by fixed records on magnetic tape and must be supported by printed documentation.

Norway

Work was initiated with a feasibility study in 1979 that resulted in a standard format being published in 1985 (Reite 1989). Called SOSI, the format was upgraded in July 1989 (Version 1.3). It is the standard format for national mapping in Norway and is maintained by the Norwegian Mapping Authority. It is expected to undergo further development as more users become experienced with it. A program library is available to aid users in interfacing with it.

South Africa

A proposed national standard for South Africa draws upon the earlier experiences of other countries (Cooper 1989). This standard is based upon a relational model and therefore offers a great deal of flexibility (Lane 1988). It attempts to cater to all forms of digital geo-referenced data, and one of its strengths appears to be in handling relations between features, and the easy addition of new relationships, when needed.

Sweden

A Research and Development Committee for Land Information Technology was established by the National Land Survey of Sweden to investigate the development of national mapping standards. A project to develop or adopt a "complete" exchange standard was launched in 1989 (Persson 1989). Its priority was to develop a standard for the description of data quality for use with an existing format and to develop a magnetic tape transfer standard for raster data. The work is being done under the auspices of the Swedish Standards Institute (SIS) and consists of a steering committee, a technical committee and five working groups, namely: GIS Terminology and Feature Classification; GPS Terminology; Digital Field Survey Systems; Raster Data Transfer Format; and Data Quality. This work is to be completed by 1991.

United Kingdom

The development of national standards in the United Kingdom has been the responsibility of the Ordnance Survey, which appears to have been more successful than many other countries. This is probably because of the stronger role the Ordnance Survey plays compared to similar organizations in other countries where many diverse views complicate and slow down the process of standards development. A simple format (DMC) was implemented in the early 1970s and found widespread use in the United Kingdom. It was upgraded in 1984 for the transfer of more sophisticated data, and renamed OSTF. Both these formats will be phased out by 1992.

In 1985 a new organization comprised of a steering committee and a working group was established to resolve issues for improving the format, dealing with data quality and data classification. This resulted in a preliminary National Transfer Format (NTF) in 1986, with a final draft published in 1987, and an extensively revised version published in 1989 (Sowton 1989). There are more than 100 registered users of NTF and, as mentioned earlier, it is also the basis of GDF, a proposed standard for road maps to be used for automobile navigation systems. NTF has four levels of transfer. They are: 0—Raster or gridded data; 1—Simple vector data; 2—Multiple attribute and quality data; 3—Topology; and, 5—A user definable data dictionary which may be based upon ISO 8211.

United States of America

Several formats have been developed in the U.S., but this discussion is limited to the major effort now taking place to develop a Spatial Data Transfer Specification (SDTS) to cover vector, raster and relational data structures. This has been a major effort which has received extensive publicity, and several other countries are considering it as a basis for their own national standard. Three related groups participated in its development: The National Committee for Digital Cartographic Standards (NCDCDS); the Standards Working Group of the Federal Interagency

Coordinating Committee on Digital Cartography (FICCDC); and the Digital Cartographic Data Standards Task Force (DCDSTF). The responsibility for finalizing the standards now rests with the U.S. Geological Survey, and an internal maintenance authority was established in 1988 to test the proposed specifications (Rossmeissl 1989). The standard is being revised as a result of the testing, and it is hoped that it will be reviewed by the National Institute of Standards and Technology (NIST) in 1990 and be subsequently issued as a Federal Information Processing Standard (FIPS).

International Activities

There are two efforts under way to actually develop international standards. These are under the auspices of the International Hydrographic Organization (IHO) and the Digital Geographic Information Working Group (DGIWG). In addition, a major project by the United States Defense Mapping Agency (DMA), in cooperation with Australia (DOD), Canada (DND), and the United Kingdom (MOD), to create a PC based vector database of the world may establish a *de facto* world standard (McKellar and Feeley 1990a) for digital map data.

International Hydrographic Organization (IHO)

The IHO is an intergovernmental consultative and technical organization of fifty-seven member states which are concerned with the collection and dissemination of information for safe marine navigation. Part of its responsibility is to look at emerging technologies on behalf of its member states, and to provide advice on how such advances could impact the charting community. In 1983 the IHO established a Committee for the Exchange of Digital Data (CEDD) to look at creating a format for the transfer between hydrographic offices of reprographic chart data. In 1985 the mandate was enlarged to include the exchange and dissemination of electronic navigational chart data. A "straw man" format, based upon DMAs Standard Linear Format (SLF), was proposed in 1987 (DX-87) and implemented by the U.S. National Ocean Service. However, the encoding was perceived as too restrictive and is now being replaced by a new format (DX-90), encoded using ISO 8211. A separate task is the preparation of a Feature/Object Catalogue which defines the (atomic) elements that will be used to code and describe navigational features. Although earlier approaches were based on a model of the navigational chart, the new standard endeavours to model reality. Tests using these standards in a small area of the North Sea are planned for late 1990. These tests are being organized by the IHO Committee on Electronic Charting (COE) and its working groups.

Digital Geographic Information Working Group

DGIWG, whose membership is made up of NATO nations, consists of a steering committee, a technical committee and a panel of experts. An exchange standard specification known as DIGEST has been produced for vector, raster and digital terrain models data. It also includes a comprehensive feature coding scheme covering land mapping, hydrographic charting and aeronautical charting. The current specifications focus on the use of physical media for exchange and has proposed that the ISO 8211 standard data descriptive file format be used for encoding the data. However, for the telecommunications of map data an alternate encoding would be more appropriate. On behalf of DGIWG, Canada (DND) has studied how to encode DIGEST or any other set of geographic data for telecommunications, and has proposed a Digital Geographic Architecture (DGA) that separates the content from the carrier (McKeller *et al.* 1990b) that would cover all media, physical and electronic, in one integrated solution. Such an approach will be important when bulk data is distributed in hard media such as CD-ROM, and when timely updates are sent electronically. The GDA approach parallels the development of the Office Document Architecture (ODA) by ISO.

Discussions are also being initiated between IHO and DGIWG in order to determine if it would be possible to consolidate the work of these two groups.

Digital Chart of the World (DCW)

Early in 1992 the DMA expects to be able to release a 20 Gbyte database of the world at a scale of 1:1,000,000. Using a topologically based vector structure, the product will be distributed on thirty-five CD-ROMs. The information will be derived by scanning 270 DMA Operational Navigation Charts (ONC), and the DCW may be based upon DGIWGs Digital Geographic Information Exchange Standards (DIGEST). The impact of such a concentrated effort (it will cost over $10,000,000) is not clear, but it is expected that the project will set standards for the production of GIS CD-ROMs and may also set standards for content structures and exchange formats for digital geographic information (McKellar and Feeley 1990a).

International Cartographic Association (ICA)

At the 14th World Conference of ICA, a Working Group on Digital Cartographic Exchange Standards was established and held its first meeting. Thirteen nations participated and set several goals for the group. The group hopes to be a focal point for information concerning data exchange development throughout the world; to exchange information and reports by member countries; to collect and distribute to working group members copies of all standards published in ICA countries; to identify research needs that arise from the standards process; and finally, to produce a monograph in 1990 on Transfer Standards for Digital Spatial Data. The working group does not intend to develop a world standard (Moellering 1989).

Conclusions

Establishing digital geographic exchange standards is a difficult task, and there is no clear indication that any of the existing standards or work in progress will reach any kind of broad acceptance. Many of the reasons as to why it is so difficult to achieve broad consensus have been given. However, the problems and requirements for data exchange are better understood now and there is a high probability that useful and broadly acceptable standards will become available by the mid-to-late 1990s.

International groups such as the IHO, DGIWG and ICA are playing important roles in developing standards and disseminating information about them. Further, because of the growing acceptance of the ISOs Open Systems model OSI by computer systems manufacturers, it is expected that in future formats used to encode the data will conform to the OSI model.

References

Benninger, B. (1988) "Hypercard model of Kingston Harbour", Prototype system developed for the Canadian Hydrographic Service, Department of Fisheries and Oceans, Ottawa.

Brüggemann, H. (1989) Unpublished Report: "Exchange formats for topographic-carto-graphic data".

CCSM (1984a) "Volume I: Data classification, quality evaluation and EDP file format", Topographical Survey Division, Energy, Mines and Resources Canada (EMR), Ottawa.

CCSM (1984b) "Volume II: Topographic codes and dictionary of topographic features", Topographical Survey Division, EMR, Ottawa.

CCSM (1986) "Proposed standards for a digital topographic model", EMR, Ottawa.

CCSM (1988) "Standard EDP file exchange format for digital topographic data", Canada Centre for Mapping, EMR, Ottawa.

CHS (1988) "Specification of the map and chart data interchange format: MACDIF, Version 2", Canadian Hydrographic Service, Department of Fisheries and Oceans, Ottawa.

CHS (1990) "Specification of the map and chart data interchange format: MACDIF, Version 2.3", Canadian Hydrographic Service, Department of Fisheries and Oceans, Ottawa.

Claussen, H., P. Heres and J. Siebold (1989) "GDF, a proposed standard for digital road maps to be used in car navigation systems", Conference Record of Papers Presented at the 1st Vehicle Navigation and Information Systems Conference, Toronto.

Cooper, A. K. (1989) "The South African standard for the exchange of digital geo-reference information", Auto-Carto 9, Baltimore.

Evangelatos, T. V., J. Yan and B. Haddon (1990) "Geomatic standards in the Federal Government", Proceedings of the Second National Conference on Geographic Information Systems, Ottawa.

ISO (1985a) "Specification for a data descriptive file for information interchange", International Standards Organization, ISO 8211-1985(E), Geneva.

ISO (1985b) "OSI—Specification of Abstract Syntax Notation One (ASN.1)", International Standards Organization, ISO 8824, Geneva.

Lane, A. (1988) "A review of a National Standard for the exchange of digital referenced information by Clarke, Cooper, Liebenberg and van Rooyen", *International Journal of Geographical Systems*, Vol. 2, No. 1.

Lindsay, G. (1989) Unpublished presentation to the ICA Working Group on Digital Cartographic Data Exchange Standards, Budapest.

McKellar, D. and J. Feeley (1990a) "The digital chart of the world project", *Proceedings of GIS90*, Ottawa.

McKellar, D., L. O'Brien and W. Lalonde (1990b) "An architecture for the exchange of geographic data", *Proceedings of GIS90*, Ottawa.

Moellering, H. (1989) Minutes of the ICA Working Group on Digital Cartographic Standards, Budapest.

O'Brien, D. (1988) "Data encapsulation—ISO 8211 and 8824/5", International Hydrographic Bureau Digital Data Seminar, Monaco.

OMNR (1989) "Technical specification of the mapping data interchange format—MDIF, Version 3, Draft 3", Ontario Ministry of Natural Resources, Toronto.

Persson, C. G. (1989) Unpublished presentation to the ICA Working Group on Digital Cartographic Data Exchange Standards, Budapest.

Rainio, A. (1989) "The joint use of geo-information in Finland—experiences of testing phase and estimations of influences", *Finnish Proceedings of 14th ICA Conference*, Budapest.

Reite, A. (1989) Unpublished presentation to the ICA Working group on Digital Cartographic Data Exchange Standards, Budapest.

Rossmeissl, H. (1989) "The spatial transfer standard: a progress report", Unpublished presentation to the ICA Working Group on Digital Cartographic Data Exchange Standards, Budapest.

Salgé, F. (1989) "Geographic, cartographic, located data exchange format—an overview of existing solutions", UDMS, Lisbon.

Sowton, M. (1989) Unpublished presentation to the ICA Working Group on Digital Cartographic Data Exchange Standards, Budapest.

Vahala, M. (1986) "Data standardization of the integrated LIS in Finland", *Proceedings Auto-Carto*, London.

Voelcker, J. (1986) "Helping computes communicate", *IEEE Spectrum*, March.

CHAPTER 9

Cartographic Data Display

TERRY A. SLOCUM and STEPHEN L. EGBERT

Department of Geography
University of Kansas
Lawrence, Kansas 66045, USA

Introduction

The present era is exciting for those interested in cartographic data display. We have entered an arena where technological developments and marketplace forces have placed powerful, but affordable, computers on our desktops and in our laboratories. These computers allow us to create map displays that we only dreamed about a short time ago. As a background to the chapter, it is useful to consider some of the changes that have led to present capabilities. These changes can be grouped into: (1) improvements in hardware, (2) improvements in software, and (3) the merging of the geographic techniques subfields of geographic information systems (GIS), remote sensing, spatial analysis (quantitative methods), and cartography.

Improvements in Hardware

One of the major changes in hardware was the movement from mainframes to microcomputers beginning in the late 1970s (see chapter by Coll). Many cartographers can probably recall working in clumsy batch-oriented environments on mainframe computers using systems such as SYMAP. With the advent of microcomputers, cartographers had the luxury of being able to create a map within a few seconds on a CRT or a few minutes on a hardcopy device without having to worry about such mundane matters as whether one's computer account would run out of money. (None of our comments should be taken to mean that mainframes have been consigned to the rubbish heap—the computational speed and power of minis, mainframes, and even supercomputers have been, and will continue to be, indispensable for addressing some of the difficult problems faced by cartographers.)

Although microcomputers such as Apple, IBM, and others freed cartographers from mainframes, the real advantage of these systems was their affordable color raster graphics capabilities. Tektronix graphics terminals and microcomputers were available to many cartographers prior to the late 1970s, but these were based on vector storage tube technology. Vector systems generally were limited to monochrome display and required erasure and a complete redrawing of the screen if any portion of an image was changed. In contrast, color raster graphics permitted the display of a variety of colors and the changing of any portion of the screen selectively, without erasing and redrawing the entire image. Being able to selectively update features in a map was essential for the full development of interactive cartography.

Another important change in hardware has been the introduction of cartographic workstations. Although their cost is still beyond the reach of many cartographers, they are exciting because of their greater graphics potential, for example the ability to examine three-dimensional images in real time. As microcomputer capabilities converge with those of workstations (Weissman 1988), many more users will have access to workstationlike capabilities.

Improvements in Software

Although improvements in software generally have not kept pace with changes in hardware, there have been a number of significant developments. One is that an interest in digital cartography no longer requires extensive knowledge of computer programming, as was the case prior to about 1980. Substantial software is now available for the cartographer wishing to design and produce maps (see chapter by Keller and Waters); moreover, no special expertise in computer programming is required to use this software.

Another significant software development is that those who create display systems now find the task simpler. Graphics primitives (e.g. MetaWindow and Halo in the IBM-compatible world) permit the execution of basic graphic functions such as shading areas with a single line of code. The projected wide availability of object-oriented programming tools and associated graphics primitives should make this task an even easier one in the future (e.g. Steiner et al. 1989).

Although skill in programming is still necessary for those creating display systems, the trend is toward software that requires less programming expertise. Probably the best example is the HyperCard (and SuperCard) software for the Apple Macintosh microcomputers (see chapter by Raveneau et al.). In discussing the advantages of HyperCard, Crane (1988, p. 40) states:

"The more control we can give to those most advanced in a subject, the better they will be able to apply the intuitions and instincts of the expert, and the more innovative and intellectually stimulating new academic materials will be."

HyperCard truly is a unique piece of software; it can function as an application program, a screen painting program, a database program, or as a complete programming language (Vaughan 1988, pp. 8–15). Programming in HyperCard is accomplished using a scripting language called Hypertalk which has a number of object-oriented features that make it easier to use than standard programming languages such as C, Pascal, or FORTRAN.

Another important software development is the ease with which graphics, text, speech, and pictures can be linked in a nonsequential fashion; authors have variously referred to this as hypermedia or interactive multimedia (Vaughan 1988; Miller 1989; Robinson 1990; Lippincott 1990). From the standpoint of cartography, this development means that maps can be linked with a wealth of other information. Developing hypermedia is partly a function of available hardware such as the videodisk, but it is discussed here because of the critical role of the software. Although hypermedia might be implemented using traditional programming languages, such as C, the ease with which HyperCard can be used makes it an ideal tool.

Merging of the Geographic Techniques Subfields

Prior to 1980 the geographic techniques subfields of GIS, remote sensing, spatial analysis, and cartography were separate entities. With the rapid growth of GIS in the 1980s, these subfields are now clearly merging (Goodchild 1988; Tomlinson 1988; see also chapter by Taylor). Within GIS, cartography is often considered the data display component. Some cartographers may feel awkward about placing cartography within the realm of GIS—they may feel that they are being engulfed by GIS and in danger of losing their identity as cartographers. Instead of worrying about whether cartography is a subset of GIS, cartographers should be taking advantage of all that GIS has to offer. GIS provides a fertile ground for linking maps with a variety of other geographic information in much the same way that HyperCard links graphics, text, and pictures. Recognizing that cartography is an important element of GIS will not prevent it from continuing to function effectively outside of GIS (cf. chapter by Taylor).

Taking Advantage of Recent Changes

Clearly, the changes discussed are exciting ones for cartographers, but has full advantage been taken of the potential these changes offer? In answering this question, it is useful to consider two potential goals. One is to use digital technology to create maps that are indistinguishable from their manually-created counterparts. The other is to use the computer tools at our disposal to create a new array of methods for displaying and analyzing spatial data. If developments in digital cartography over the last 30 years are examined, it is apparent that many efforts have been oriented toward the first goal, i.e. producing maps equivalent to manual products. Although the emulation of traditional manual cartography is a desirable and necessary first goal, it is a limited one. In addition to emulating traditional printed maps, cartographers need to explore new ways in which modern computer technology can be used to display and analyze spatial data.

In reviewing the thoughts of others on this subject, it is clear that these ideas are not new. As early as 1965, Tobler (p. 37) stated:

> "One of the important implications of automated methods in cartography is that they disrupt the traditional attitudes and force a re-examination of many of the conventional cartographic procedures."

In research published in 1979, Anderson and Shapiro (p. 21) expanded on this theme.

> "Because of their fundamental differences, the design of electronic maps should not mimic that of paper maps. Each cartographic decision or design feature must be reconsidered based on its underlying purpose."

Anderson and Shapiro listed a number of advantages that interactive electronic maps had over traditional paper maps, including the capability to: (1) avoid clutter by displaying selected information, (2) derive information about an object pointed to, and (3) focus attention through blinking or changes in video intensity.

At about the same time that Anderson and Shapiro were doing their research, Moellering (1978, 1980) was emphasizing how digital technology could be used to explore three-dimensional displays in real time. Moellering's approach was clearly something that manual cartography could not provide, but it was, as he later said, ". . . regarded by most cartographers as being rather exotic . . ." (Moellering 1984, p. 127).

In more recent years, others have commented on the untapped potential of digital map displays. For example, Taylor (1985, p. 20) has promoted the notion of a "new cartography", with regard to which he states:

> "In design terms it includes elements such as the extensive use of colour, sequencing and animation for images on home TV and microcomputer screens that do not occur in conventional mapping."

Interestingly, Goodchild (1988, pp. 312, 317–318) also has promoted the notion of a "new cartography". Whereas Taylor defines a new cartography in terms of technology, design, theory, and education, Goodchild focuses on the display capabilities of the new technology. He states:

> "The most straightforward objective of digital technology is the emulation of manual methods, to the point where the two products are indistinguishable. . . . [However,] it is intuitively unreasonable that a technology optimized under the narrow constraints of pen and paper would turn out to be indistinguishable from one optimized under the much broader constraints of digital technology. . . . [The] most significant impact of digital

methods on cartography will occur when the field evolves to fit the constraints of the new technology and moves beyond its traditional limitations."

Goodchild offers scene generation and the time dimension as two areas in which the new technology could be particularly useful.

Marble (1987) also has suggested that we need to develop a broader view of digital cartography. He points out that one of the keys to this broader view is a clear separation of data storage and data display. In a fashion similar to Goodchild, Marble stresses the capability of modern technology to handle the time component through the use of dynamic displays.[1]

With these thoughts in mind, the focus of this chapter is on recent developments (primarily over the last 10 years) that go beyond the display methods that traditional manual cartography has produced. The discussion is not meant to be exhaustive, but it is hoped that it will provide readers with some sense of what has been done and what needs to be done in the future.[2]

Recent Developments in Data Display

To simplify the following discussion, recent developments in data display have been divided into three categories of mapping: static, interactive, and animated. The term static mapping refers to developments that produce digital maps analogous to traditional printed maps. In contrast to static mapping, interactive mapping is characterized by the ease with which individual maps can be changed or the ease with which additional maps can be displayed on a CRT. Finally, the term animated mapping refers to map displays characterized by continuously changing content or form.

Such a categorization is not without problems. For example, certain interactive mapping systems provide animation as one distinct function, while some animated mapping systems permit the user to have control over (and thus interact with) the map display by controlling the speed of animation. Placing a mapping system in one of these categories is based on the primary focus of the display method.

Several comments are in order regarding the following discussion. First, in considering the references selected, the reader should realize that the focus is on the final map display, rather than on the analytical processes preceding the display. For example, the problem of spatial interpolation (e.g. Lam 1983) is not dealt with here; nor are problems of generalization caused by scale change prior to display generation (e.g. Nickerson 1988).

Second, the reader should assume that the data display methods discussed are appropriate for a microcomputer (or workstation) unless stated otherwise. A distinction is not made between the microcomputer and workstation because the capabilities of these are merging.

---

[1] Monmonier (1982a), Morrison (1986), and Muller (1989) have also addressed these issues.
[2] For discussions of earlier developments in data display, see Dudycha (1981), Monmonier (1982b), and Carter (1984).

Third, note that illustrations are presented for only a limited number of the display methods. Although it generally is feasible to illustrate static maps, it is often difficult or impossible to illustrate interactive and animated maps on the printed page. As a final comment, note that space does not permit a complete discussion of the role of the map user and relevant experimental research; rather the focus is on the various types of displays that can be generated.[3]

Static Mapping

Univariate choropleth maps

Univariate choropleth maps have been the most popular form of thematic display in digital cartography, and this has been reflected in the variety of work that has been done with them. The landmark study in this respect was Tobler's (1973) development of the unclassed choropleth map. In manual cartography, the convention was to class the data and use a limited number of shades (typically five to seven) to represent the classes. Tobler showed that the line plotter could create a virtually infinite set of shades by simply varying the spacing between solid cross-hatched lines. The resulting unclassed shadings had the added advantages of eliminating the classification error introduced by grouping unlike values.

Again using a line plotter, Brassel and Utano (1979) made a number of modifications to Tobler's approach, including an improved legend and symbol scheme, a method for focusing on one portion of the data (termed quasi-continuous mapping), and the inclusion of a histogram of the data. In quasi-continuous mapping, one could focus on a portion of the data by applying continuous shading to only that portion. For example, for a data set ranging from 0 to 100, values from 50 to 100 would be black while those from 0 to 50 woud range from white to black. Through the inclusion of the histogram, Brassel and Utano illustrated the importance of combining maps and graphs and of looking at data from a variety of perspectives, something that we might expect to find in a modern multimedia system.

Following up on Tobler's idea, Sibert (1980) implemented the concept of unclassed mapping on CRTs. Today, many users can create unclassed maps because of the decreasing cost of color boards capable of displaying 256 colors at one time out of a much larger palette. Although unclassed maps can now easily be created, they are not used frequently. One reason is that much of the software for choropleth mapping has been based on the IBM Enhanced Graphics Adapter (EGA), which cannot be used to create an unclassed map due to its small palette and limited number (16) of display

---

[3] Kimerling (1989) provides a summary of experimental research in cartography. Other articles particularly pertinent from the map use standpoint include those by van Elzakker and Ormeling (1984) and Robertson (1988).

colors. Unclassed maps also are not used frequently because cartographers have never agreed that they are desirable (see Dobson 1980a and Muller 1980 for an example of the debate).

Other developments for univariate choropleth maps have dealt with methods for creating them with various output devices. Groop and Smith (1982b) and Plumb and Slocum (1986) have described methods for creating choropleth maps on dot-matrix printers. In a similar fashion, Goulette (1985) and Gilmartin (1987, 1988) have explored the creation of choropleth symbols on CRTs with limited color capabilities.

Bivariate choropleth maps

Bivariate choropleth maps were developed by the U.S. Bureau of the Census in association with its improved digital capabilities. A bivariate map was constructed essentially by overlaying two univariate choropleth maps. Although the bivariate map could have been produced by conventional manual methods, the process was accelerated through the use of a computer-output-on-microfilm (COM) device (Meyer et al. 1975).

Recent advancements in the area of bivariate mapping include the work of Eyton (1984), Carstensen (1982), and Lavin and Archer (1984). Eyton's major contributions include the development of a complementary-color scheme and the use of the reduced major axis and associated bivariate normal distribution. In the complementary-color scheme, he represents separate variables by cyan and red; when overlaid, these two colors create gray tones wherever they are combined in equal amounts. This approach leads to a map that is clearly more logical than one based on the original color scheme used by the Bureau of the Census.

Eyton's reduced major axis approach facilitates map interpretation by considering the statistical relation between the variables when assigning colors. In addition to the above contributions, Eyton also showed how the complementary-color scheme could be used to create an unclassed bivariate map. He conceded, however, that ". . . the larger number of color tones inherent in the map precludes easy map analysis" (Eyton 1984, p. 489).

At about the same time Eyton presented his ideas, Carstensen (1982) and Lavin and Archer (1984) demonstrated how the line plotter could create unclassed bivariate maps; in their approach the two variables were represented respectively by horizontal and vertical lines of differing spacing (Fig. 9.1).

Both Carstensen and Lavin and Archer felt that the quantitative changes in line spacing created on the line plotter would be more logical for the map reader than the original Bureau of the Census color scheme. Furthermore, in agreement with Tobler, they felt that the unclassed map would eliminate the quantization error caused by classification.[4]"

[4] For other work in the context of bivariate mapping, see Smith (1977), Monmonier (1978), Carstensen (1986a, 1986b), and Hartnett (1987).

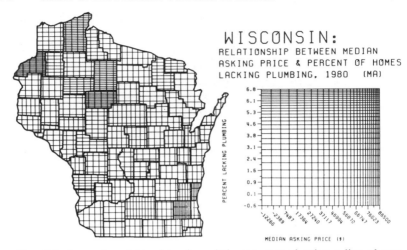

FIG. 9.1. An unclassed bivariate choropleth map created using a line plotter. Depicted is the relationship between median asking price for homes and percent of homes lacking plumbing – Wisconsin, 1980 (from Carstensen 1986a, p. 36).

With regard to bivariate choropleth maps, an interesting study would be to compare the effectiveness of Eyton's complementary-color method with that of the line plotter approach for both classed and unclassed maps. Such a study is essential because in the future Eyton's approach may well be preferred due to its fine texture and the ease with which colors can be generated on the CRT.

Displaying continuous data

A number of novel approaches have been developed for portraying continuous geographical data, largely because of the following problems with the most conventional method for continuous data, the isoline map:

> "One is the fundamental inconsistency in using discrete symbols to represent continuous data. A second, and, perhaps a consequence of the first point, is the likelihood that among our symbolic forms isolines are the most taxing to interpret and understand" (Lavin 1986, p. 140 in referencing Dent 1985).

In one approach, known as the dot matrix method, Groop and Smith (1982a) first divide the map area into a grid of very small equal-shaped invisible cells (e.g. hexagons). Each cell is then filled with a shaded symbol having the same shape as the cell, the size of the symbol being proportional to the magnitude of the datum at that location. After photographic reduction the map has a "... continuous gray tone change [that] is consistent with the conceptual surface it represents" (Groop and Smith 1982a, p. 129) (Fig. 9.2).

In a second approach, known as dot-density shading, Lavin (1986) first divides the map area into a grid of small equal-sized square cells. The

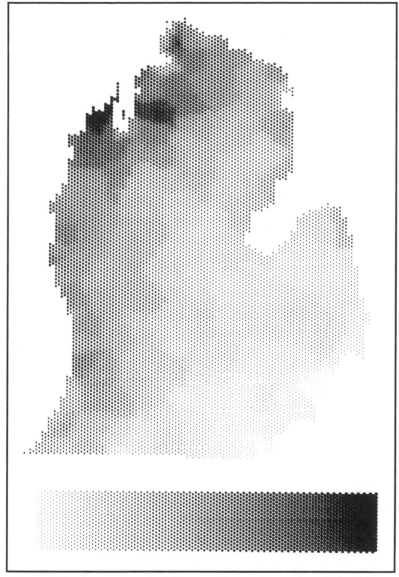

FIG. 9.2. The dot matrix method for portraying continuous data. Shown here are snowfall totals for Michigan's lower peninsula, 1978–79 (from Groop and Smith 1982a, p. 128).

number of dots placed in a cell is then made proportional to the magnitude of the datum at the center of a cell and the dots are placed within a cell using a stratified-random procedure. The resulting map looks like a traditional dot map, but it is based on continuous rather than discrete data.

Although the Groop-Smith and Lavin approaches provide an interesting picture of continuous data, they share a weakness of not providing specific information, such as the value at a location. In this vein Lavin argues that his method be used as a complement to isoline mapping. He states:

> "People who make maps tend to have a uni-map mentality; they search for the best single symbolization to use on a map. In actuality, a geographical topic may be better illuminated by multi-map display, using more than one form of symbolization for a single data set" (Lavin 1986, p. 148).

Lavin's notion fits well within interactive systems in which data are explored from a number of different perspectives.

Lavin and Cerveny (1987) are responsible for another innovative approach, termed unit-vector density mapping. Their approach is a modification of Lavin's dot-density procedure, with line segments used instead of dots. If line segments are used alone, the method can show elevation and slope direction, and if arrowheads are appended to line segments, the method can, in theory, show both speed and direction of wind flow. The very small arrows used in the latter case may, however, make the determination of wind direction difficult (Fig. 9.3).

One limitation of all of these methods for displaying continuous data is the length of time required to create maps. For instance, Lavin and Cerveny (1987, p. 140) indicate that it took anywhere from 10 to 35 minutes to produce the unit-vector density maps. Another problem is that in their present form the methods are limited to very high-resolution hardcopy devices. In this respect Groop and Smith's dot-matrix method could be modified by using the pixel on a CRT as the basic cell; this cell could then be lit with an intensity proportional to the datum at that location. In fact Groop and Harman (1988) have used this concept to portray distributions (such as climatic regions) characterized by transitional boundaries.

Three-dimensional maps

The major development in static three-dimensional maps has been the ability to overlay thematic data such as land use on top of topographic data (e.g. Junkin 1982; White 1985; Eyton 1986; Grogan 1986). Although such maps can be created by manual methods, the process requires the hand of a skilled cartographer and a great deal of time (Jenks 1989). White's approach is attractive because of the apparent ease of computation and its appropriateness for CRT display. Eyton describes an interesting procedure for portraying the effect of elevation and latitude on mean annual temperature.

Other research in static three-dimensional mapping includes work by Brassel and Kiriakakis (1983) and Clarke (1988). In the latter case Clarke has developed a sophisticated method for simulating topographic relief based on Fourier and fractal methods. Although geographers obviously have a great

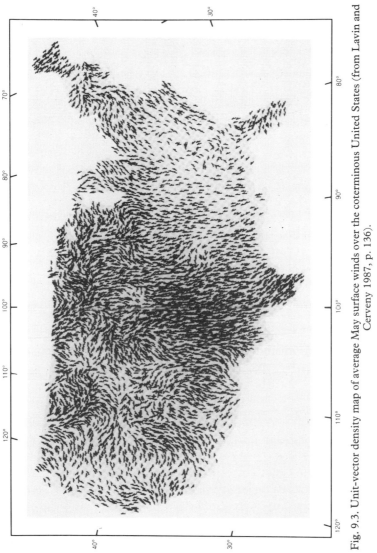

Fig. 9.3. Unit-vector density map of average May surface winds over the coterminous United States (from Lavin and Cerveny 1987, p. 136).

deal of digital topographic data available, Clarke argues that there are a number of instances in which simulated terrain would be useful. For example, he notes that Morrison (1986)

". . . has suggested that simulated topography could be used to fill in background information on thematic maps where the need is merely to convey impressions" (Clarke 1988, p. 173).[5]

Other significant developments

Some examples of other developments in static mapping include the work of Cuff *et al.* (1984), Yoeli (1985a), and Fairchild (1987). Cuff *et al.* (1984) developed the nested value-by-area cartogram for displaying land use and other proportional data associated with areal units (e.g. counties). In their approach, cartograms of each county, scaled in proportion to the data, are placed within their respective county boundaries on a base map (Fig. 9.4). This procedure is a modification of original developments with non-contiguous cartograms (Olson 1976) and takes advantage of the digital outlines available for each county.

A great deal of work in data display has been done by Yoeli (e.g. 1983, 1984, 1985a, 1985b, 1986). Although much of this work is an attempt to emulate traditional cartography in a digital environment, his work on the construction of intervisibility maps (Yoeli 1985a) goes beyond emulation. With regard to manually created intervisibility maps, he states:

". . . the accuracy of the result depends largely on the scale of the contour map used, the vertical interval of the contours and their accuracy, the density of the profiles constructed and the accuracy of their drawing . . ." (Yoeli 1985a, p. 88).

Yoeli argues that many of these problems can be avoided using the computer.

Interactive Mapping

Choropleth maps

In a fashion similar to static mapping, a considerable amount of work has been done with choropleth maps in an interactive environment. Turner and Moellering (1979) were responsible for some of the earliest research in this area. The objective of their interactive system was to make the production of choropleth maps

". . . as easy as possible, and yet at the same time offer the cartographer as many options as might be useful in making the maps" (Turner and Moellering 1979, p. 256).

---

[5] Clarke's work is a data display method that is difficult to classify. If the image could be created fast enough it would clearly be useful in an interactive or animated environment; since he indicated that his terrain models could be generated in "under 2 minutes", it has been placed in the static mapping category.

PROPORTION OF AREA IN
DESIGNATED LAND USE

LAND USES

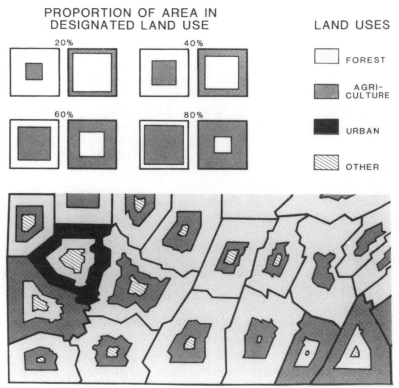

FIG. 9.4. Nested value-by-area cartogram showing southwestern Pennsylvania land use by county, with cartograms adjusted to consider perception (from Cuff *et al.* 1984, p. 6).

Some examples of options included: (1) the choice of a variety of classification procedures (or, alternatively, an unclassed map); (2) a method for moving cartographic objects by pointing to screen locations; and (3) the display of a histogram to determine whether the data need to be transformed. In a sense, Turner and Moellering's approach can be viewed as a transition between static and fully interactive mapping systems. Their system was clearly interactive in that great flexibility in design was permitted; one goal, however, was to produce a static hardcopy map.

The DIDS system, or Domestic Information Display System (Dalton *et al.* 1979) has been viewed by many as one of the first mapping packages to take significant advantage of computer graphics capabilities. When initially conceived, this system was designed as a decision support tool for the U.S. federal government. To achieve this purpose, DIDS permitted the display of county-level maps of the United States (3000+ counties) within a few seconds. A key feature of the system was a zoom function for examining subregions in greater detail. As Moellering (1984, p. 128) indicates, this was a "... very slick and flashy system ...". Unfortunately, DIDS was based on hardware to which most cartographers did not have access. Moreover, the system was later discontinued (Cowen 1984).

More recently, Slocum et al. (1988) have developed a system to assist users in acquiring information from a choropleth map. For the purpose of assisting the user, a hierarchical menu system is available that allows the examination of a map in a variety of ways. At the top level of the hierarchy, users select either general or specific information, that is, information about patterns or specific areas, respectively. When the general information option is selected, users can sequence classes, compare classes, or view single classes. By choosing the specific option, users can perform functions such as comparing two or more areas or counting the number of areas comprising a class.

The sequencing concept included within the general option in Slocum's system originally was developed by Taylor (1984, 1987). In sequencing, classes are displayed in order from, say, low to high. In theory, sequencing should enhance interpretation of the map by both providing the reader with "chunks" of information (Taylor 1987) and emphasizing the ordering of the data from low to high. Although sequencing is an interesting attempt to take advantage of automated capabilities, it should be realized that research by Taylor (1987) and work that the authors are currently undertaking does not fully support this idea. In the long run, sequencing may prove more useful with other types of thematic maps.

Slocum's concept of an information system for choropleth maps raises some challenging questions concerning the manner in which map readers will make use of such systems, and the level of training required to do so. Traditionally, readers have been given a single map characterizing a spatial distribution; generally, this has been true whether the map has been produced in hardcopy form or on a CRT. But, if readers have the ability to interact with the map and to explore the data set from a variety of spatial perspectives, how will they go about doing this? Would such a capability be of interest only to sophisticated users? Is there a need to train naive users who wish to use such systems?

Slocum's research also stimulates anew the issue of classed versus unclassed choropleth maps. Dobson (1988) argues that if one can point to an areal unit and determine its value, there may be little reason to class the data. It is particularly noteworthy that Dobson has made this argument, since he originally was the major opponent of the unclassed approach (Dobson 1973, 1980a, 1980b).

Another piece of research that clearly benefits the user is Olson et al.'s (1989) work in progress on maps for the color deficient. Although it has wider application than choropleth maps, it is mentioned here because of the clear user orientation. Olson's objective is to determine color schemes that are effective for the color deficient. Once such schemes are found, options could be included in software packages to display color schemes that are appropriate for the color deficient; obviously, such an approach would be too costly in a manual environment.

A fascinating example of the use of choropleth maps in an interactive environment is the work that Miller (1988) has done with the Great American History Machine (GAHM). Miller created GAHM to encourage undergraduates to think like professional historians. Traditionally, this was done through extensive reading; GAHM is viewed as a shortcut to this lengthy process. In GAHM, students have access to the entire county-level U.S. census returns from 1840 to the present, along with presidential election returns. Using advanced workstations such as the Sun 3/50, students can analyze these data sets via choropleth maps they create themselves. In addition to analyzing a particular data set with a single choropleth map, students can compare two variables on separate maps or compare a variable with a map of physical features, using a split-screen feature (Fig. 9.5).

In evaluating the potential of a system like GAHM, it is natural to ask why geographers (for whom maps are a fundamental tool) haven't developed similar systems. In a fashion similar to historians, geographers should be interested in encouraging undergraduates to think like professional geographers. To some extent one can argue that current GIS software serves this function. There are two problems with this argument, however. First, current software is oriented toward the physical side of the discipline; many in human geography, particularly in cultural and historical geography, may be unaware of its potential benefits. Second, much of the software is characterized by terse command-driven interfaces that make the direct creation of simple maps difficult, especially for beginning undergraduates. Hopefully, an examination of Miller's work will give geographers the impetus to develop similar systems.[6]

Electronic atlases

One of the difficulties in discussing electronic atlases is defining them. Conceivably, an electronic atlas can range from an automated version of a book to a complete information system in which map queries and analyses are possible. An example tending toward the former is Smith's (1987) Electronic Atlas of Arkansas. In this atlas, users can view a large variety of maps for the state of Arkansas simply by selecting them from a series of menus; text associated with a map is available by toggling to another screen. This atlas must be viewed as a significant accomplishment because the graphic quality is high given the limited resources that were used to create it. In particular, the system was created for an IBM/PC, EGA-compatible

---

[6] For work on interactive bivariate choropleth maps, see Cowen et al. (1984).

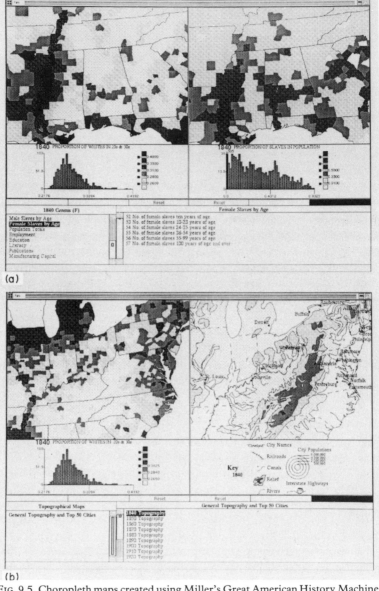

FIG. 9.5. Choropleth maps created using Miller's Great American History Machine. In (a) two variables (proportion of whites in their 20s and 30s and proportion of slaves in the population) for the same time period are compared, while in (b) a single variable (proportion of whites in their 20s and 30s, 1840) is compared with a map of physical features (from Miller 1988, p. 29).

system using EGA Paint, with Smith doing much of the design work himself.

An example of an atlas tending toward a complete information system is the Electronic Atlas of Canada (Siekierska 1984; Siekierska and Palko 1986).

This atlas is "... a microcosm of a National Atlas Information System that is being developed to support the National Atlas of Canada program ..." (Siekierska and Palko 1986, p. 409). It permits users to: (1) overlay pre-existing files to focus on a particular topic; (2) design their own maps; (3) zoom in on areas of interest; (4) highlight small areas through a blink function; (5) query geographic areas for related information; and (6) perform simulation modelling.

In between these two extremes, there are electronic atlases with various capabilities. Waters and De Leeuw (1987) describe the ATLAS program that allows the user to locate countries or cities, find facts about countries, or compute distance between places. A more recent example is the Electromap World Atlas, which Smith also was a leader in developing. This impressive package permits the examination of over 200 full-color thematic, relief, and reference maps of the world and individual countries. Text is also available describing the geography, people, government, economy, and communications of individual countries. Moreover, all of this information is available via easy-to-use drop-down menus on an IBM-compatible system having an EGA color board. Raveneau and his colleagues provide an interesting discussion and examples of electronic atlases in this volume.

Multimedia systems

Within the field of geography, the Domesday Project (Openshaw and Mounsey 1987) is clearly the star in the multimedia realm. This project was developed by BBC Enterprises Ltd. to present a contemporary view of the United Kingdom. The key to the project is the use of videodisks. One videodisk, termed the community disk, contains information collected by schools, community groups, and individuals. The other, the national disk, contains data collected from both government and quasi-government sources. Together, the disks provide the user with approximately 250 megabytes of digital data, 50,000 photographs, and 20 million words of text.

All of these data can be accessed easily through an interactive system to learn about the United Kingdom. Features of the system include the ability to: (1) move through a hierarchy of scales; (2) display a digital map on top of an analog base map; (3) retrieve data by pointing to areas; and (4) perform some simple overlay functions. Openshaw and Mounsey (1987, p. 178) indicate that 10-year-old children were demonstrating the system to the Prime Minister after being given only two hours training. Moreover, the system is very affordable; in 1987 it was estimated that a microcomputer-based system would cost from £2000 to £4000.

Because maps are frequently a component of almost any multimedia system, geographers also need to be aware of related advancements in other disciplines. For example, in an article dealing largely with the study of Greek, Crane (1988, p. 9) describes how HyperCard can be used to link an historical text to maps.

> "A structure of this type allows the user to approach the atlas in an entirely different way. Instead of simply reading the sequence of events in a text, she could flip through a sequence of maps and see geopolitical events on separate maps that reflected the actual chronology of the Persian Wars. Each individual map can further be linked to the relevant section of an ancient history text, so that the user can move immediately from the map to a detailed description of the event portrayed on the map."

Other significant developments

Other developments in the area of interactive systems include the work of Brown and Moellering (1982), Lai (1985), Poiker and Griswold (1985), Faintich (1986), Allard and Hodgson (1987), Robb (1987), Tobler (1987), and Egenhofer and Frank (1988). Since space does not permit a complete discussion of all of these, some key elements of several of these are discussed.

In response to the perceived limitations of static displays, Poiker and Griswold (1985) developed an interactive system for analyzing topographic data. They indicated that an interactive system should have the following functions:

(a) Shade all those areas that have steep slopes.
(b) Draw another data set (e.g. roads) on the surface, only showing the visible portions, of course.
(c) Point at the screen . . . and request the position . . . of the point that is indicated.
(d) Draw on the surface . . . additional graphic information.
(e) Slowly rotate the surface without recomputing the visibility of every element" (Poiker and Griswold 1985, pp. 409–410).

Their paper describes a methodology for handling these functions.

Allard and Hodgson (1987) implemented an interactive graphics method for mapping location-allocation solutions on an Intergraph workstation. A key advantage of using Intergraph is the capability to display any of 63 different design levels. Allard and Hodgson took advantage of this capability by assigning different ranges of allocation flow level to different levels of the design file. In so doing, they permitted themselves the flexibility to modify the amount of line clutter resulting from different allocation flow levels. Although some cartographers will argue that this concept is limited to the expensive Intergraph workstation, basic CAD packages frequently include this capability. One wonders why more cartographers are not making use of this basic approach.

Tobler (1987) has taken advantage of computer capabilities to develop innovative methods for mapping migration (Fig. 9.6). While he does not stress the interactive capability of the software nor does he specify exactly how quickly a map can be produced, it is implicit in the article and in Marble's (1987) introduction that the system has interactive capability.

(a)

(b)

FIG. 9.6. Automated methods for mapping migration. Figure (a) illustrates the gross migration to and from California from 1965–70. Figure (b) illustrates two methods for portraying net migration. In the first case migration arrows are shown between origin and destination, while in the second the arrows are rerouted to pass through adjacent places (from Tobler 1987, pp. 160–162).

Although his article focuses on analytical and symbolic questions related largely to migration mapping, Tobler (1987, p. 160) makes the following comment regarding error in displays of spatial data:

> "Ideally we would also like the map to provide an impression of the lack of precision in the data, which often come from a sample and always contain errors. For this purpose a degree of fuzziness can be obtained on a raster display screen by defocusing under program control; alternately one can use controlled variation in the contents of the refresh buffer to yield a visual instability in the displayed map. . . ."

The notion that data contain error is a frequently ignored one in cartography. Clearly, interactive graphics allow one to portray error and emphasize it to whatever degree desired.

Animated Mapping

Three-dimensional maps

Animated three-dimensional maps are an obvious solution to the hidden surface problem characteristic of static three-dimensional maps. Early developments in animated three-dimensional maps were limited to two significant accomplishments. The first of these involved the implementation of a hologram by Dutton (1979). The following quotation taken from a comprehensive review of graphical data analysis by Wainer and Thissen ilustrates the excitement that holograms might generate.

"An extraordinary tour de force of twentieth century display technology was developed by Geoffry Dutton . . . This display was a four-dimensional statistical graphic called 'Manifested Destiny' . . . The data . . . were the population distributions of the United States from 1790 until 1970. . . . The hologram image floats inside of what looks like a lampshade and turns before the viewer. As it turns one sees the mountains of population grow westward. The effect is very dramatic indeed" (Wainer and Thissen 1981. p. 232).

If holograms can generate this kind of excitement, they certainly have potential for cartography. At present, though, the lack of associated hardware and software make them but a curiosity.

The second early development in three-dimensional mapping was Moellering's (1980) research with real-time maps, as mentioned in the introduction. In Moellering's approach, a user could examine a three-dimensional surface from any vantage point very rapidly (in as little as a second). Alternatively, a video could easily be created that, when played back, would give the appearance of continuous change. Anyone who saw the video produced by Moellering (1978) had to be impressed with the potential of such real-time animation.

Unfortunately, for a number of years cartographers were not able to use Moellering's ideas because, as with holograms, they did not have the necessary sophisticated hardware and software. Today, this is becoming less true. Graphics workstations are now available that provide the user with the capability to explore three-dimensional surfaces in real time. A particularly intriguing example is the capability of the Tektronix 4200 series terminals and 4300 series workstations to create three-dimensional images using the principle of stereoscopics, an approach that Moellering (1989a) is using in a real-time satellite mapping grant with NASA. In using the Tektronix, two images are created at the very fast rate of 120 hertz, noninterlaced. The images are passed through a liquid polarizer and viewed with special radially polarized glasses (Moellering 1989a). Admittedly, the limitation of this device to a single manufacturer and the expense involved make this method just another curiosity for most cartographers. In spite of this limitation, the clarity with which the Tektronix produces a three-dimensional image and the fact that the image can be rotated in real time make this an exciting possibility for the future.

It should be pointed out that in addition to making use of the Tektronix in his NASA grant, Moellering (1989b) has also developed a logical and effective procedure for overlaying a satellite variable on top of topographic data. Based on work by Robertson and O'Callaghan (1985), the method involves manipulating the HLS color space in which hue-saturation and lightness (intensity) are orthogonal; the satellite variable is shown by a hue-saturation combination while the topographic data are shown by lightness. Kimerling and Moellering (1989) have also developed an effective method for displaying slope-aspect data based on opponent process color theory.

Currently, three-dimensional (3D) maps are generating excitement in the field of GIS (Lang 1989). Past 3D approaches are now referred to as "$2\frac{1}{2}$D" because they could be viewed in perspective, but could not portray a solid model (Lang 1989, p. 41; Smith and Paradis, 1989 pp. 324–325). True 3D display in the earth sciences allows one to

> ". . . see the source data, to select different iso-surface levels, to assign colors to these levels, to slice edges from the model, to peel off iso-surfaces, to rotate around the display and to zoom in and out" (Smith and Paradis 1989, p. 333) (Fig. 9.7).

McLaren (1989) summarizes a number of interesting 3D GIS applications outside of the earth sciences. For example, with respect to road/traffic engineering he describes the need to visualize the perceptual problems encountered by drivers, such as ". . . line of sight difficulties, incorrectly positioned street furniture or poor lane markings" (McLaren, 1989 p. 12).

A common thread running through many of the 3D GIS applications is the need to create realistic scenes. Since realistic scene generation is computationally intensive, it will be some time before users see full 3D GIS capabilities in a microcomputer environment. Still, the excitement that 3D GIS seems to be generating bears close scrutiny by cartographers.

Spatial-temporal data

The display of spatial-temporal (geographic time-series) data is an obvious application area for animated maps. Dutton's hologram (1979) of population change in the U.S. over time (described above) is one example. In addition to allowing one to explore a three-dimensional surface, Moellering's (1980) work (previously described) also was useful for the analysis of spatial-temporal data. In a fashion similar to Dutton, Moellering portrayed population change using three-dimensional prisms, with the map being rotated for every 20-year period to display the entire surface. In another example, Moellering displayed the diffusion of the innovation of farm tractors in the midcontinental U.S.

In more recent work Calkins (1984) has promoted the use of color to display spatial-temporal data. Unfortunately, his approach is difficult to interpret because of the lack of illustrations, a common problem in a world based on the printed page. In an oral presentation Monmonier (1989b) has provided a thorough overview of spatial-temporal data display; unfortunately, he has not as yet developed a written version of what appears to be a very interesting study.

Those wishing to develop animated maps of spatial-temporal data should realize that for certain applications the process is quite simple. To illustrate, let us say that one wished to show changes in the proportion of land in farms for each of the 105 counties of Kansas over a 50-year period. To accomplish this, one would first assign each of the 105 counties to a particular color index in a 256-color look-up table. To change the year-to-year map values, one would simply replace the RGB values for the respective indexes in the look-up table (O'Callaghan and Simons 1984). Since an entire look-up table

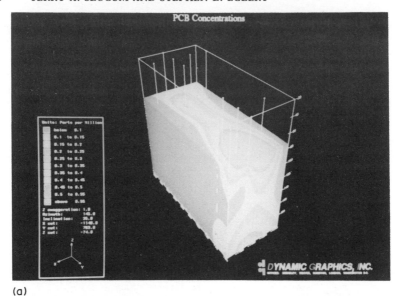

(a)

(b)

FIG. 9.7. Displays illustrating true 3D capability. Shown are two views of a zone of hypothetical subsurface PCB contamination: (a) a sliced cube, (b) an iso-surface display (from Smith and Paradis 1989, pp. 332–333).

can be replaced in less than one second, the result is the equivalent of a movie on the CRT screen. This capability is easily obtained on a microcomputer system with the appropriate color hardware. Real-time movies also can be created for applications that do not take advantage of color look-up tables, but if one goes beyond two-dimensional displays the power of graphics workstations likely will be required.

Displays for vehicular navigation

One of the fairly recent developments in digital cartography has been the capability to provide automated navigation assistance to drivers (see chapter by Claussen). This topic is relevant to animation because of the need to provide the driver with a continuous update of location.

McGranaghan *et al.* (1987) raise a number of interesting issues regarding displays for vehicle navigation. Probably the most important is whether a display is appropriate to use at all. They state:

> ". . . both theory (Kuipers) and empirical results (Streeter and others) from cognitive science suggest that a verbal-procedural presentation of such information may be more effective" (McGranaghan *et al.* 1987, p. 136).

Even if a display is used, they suggest that simple arrows could be used in place of a traditional map. Clearly, this is a case in which computer technology may change our views of traditional cartography.

Advances in related disciplines

Cartographers need to pay close attention to developments in animation that are taking place in the related disciplines of statistics and computer science. A summary of much of the work in statistics is available in a book entitled *Dynamic Graphics for Statistics* (Cleveland and McGill 1988). As an example, consider the technique of brushing scatterplots described by Becker and Cleveland (1988). In this technique a matrix of scatterplots between each pair of variables of interest is displayed on a CRT. A "brush" or rectangle is then superimposed on the CRT and moved by the user to different positions using a mouse. As the brush is moved within one scatterplot, dots within the brush are also highlighted within other scatterplots, thus enabling one to determine ". . . how a group of observations in one bivariate attribute space arrange themselves in another bivariate attribute space" (Monmonier 1989a, p. 82). Brushing can also provide capabilities beyond highlighting.

Monmonier (1989a) has illustrated aptly how the statistical brush can be modified for geographical purposes. One modification is to add a map to the initial display and highlight areas on the map that correspond to dots within the brush (Fig. 9.8).

Another modification is to query the map as opposed to the scatterplots, an approach Monmonier terms "geographic brushing". In its simplest form geographic brushing would involve highlighting particular areas on the map, an approach similar to Slocum's (1988) extraction of specific information.

Within the field of computer science, the concept of visualization has been

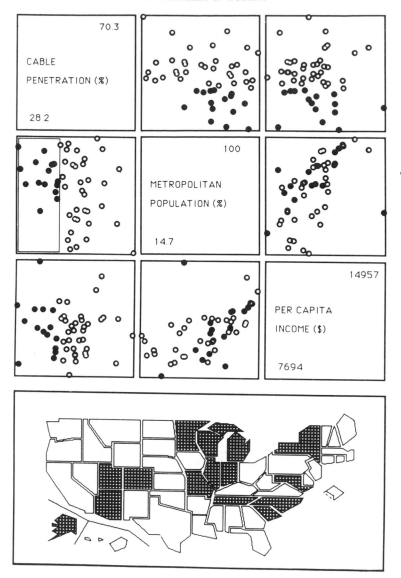

FIG. 9.8. Modifying the statistical brush for geographic purposes. The rectangular brush can be seen in the scatterplot in row 2, column 1, encompassing the fifteen states with the lowest percentages of cable penetration. The states are highlighted in all six scatterplots and on the map beneath them (from Monmonier 1989, p. 83).

a major development. The objective of visualization ". . . is to leverage existing scientific methods by providing new . . . insight through visual methods" (McCormick *et al.* 1987, p. 3). There are two ways for cartographers to view visualization. One is to say that the concept is nothing new; we have always visualized geographic problems through the use of maps. The other is to say that we admit having experience with visualization, but that we wish to take advantage of new approaches to visualization. Clearly, the latter is a more fruitful avenue to take.

The first major publication dealing with visualization was a National Science Foundation report printed in its entirety in a special *Computer Graphics* issue entitled *Visualization in Scientific Computing* (McCormick *et al.* 1987). Included with this report was a videotape illustrating a variety of sophisticated approaches for visualizing scientific problems. Several of these approaches are pertinent to cartography, two examples of which are described here. The first, entitled *L.A. the Movie*, was developed by the Jet Propulsion Laboratory (JPL). In this case a flyover of the Los Angeles, California area was created by combining a Landsat image with digital elevation data. The viewer has an inkling that something novel has been developed when the narrator begins ". . . we are now travelling towards the Pacific Ocean at approximately 200,000 miles per hour". Although the video does not quite live up to this initial billing, it certainly holds the attention; in short, one is constantly struck by the sense of flying over the landscape.

In another example (entitled *Scientific Data Visualization*, by JPL), cloud cover over the earth's surface is shown by three-dimensional cloud-like features with an overlay of continental boundaries. As the cloud cover changes over time, the three-dimensional features undulate. Unfortunately, when the entire earth is shown, the result is a bit like a choppy sea, since the viewer has no control over the rapid rate of change. When the video zooms in on a smaller region, however, visualizing the degree of cloudiness is easier.

One problem shared by a number of the visualization examples is the computational power that is needed to generate them. For example, 130 hours of CPU time on a VAX 8600 mainframe were required to generate *L.A. the Movie*. As a result, most of the examples are beyond the realm of the typical user. For many other visualization problems, however, including those that would be addressed by geographers and cartographers, workstations or the most recent generation of microcomputers would provide sufficient power and display capabilities.

Another problem with the visualization examples is the inability to interact with the video. For instance, the viewer watching *L.A. the Movie* might want to slow the flyover and control its direction. Allowing such interaction clearly would require more than a home television set and a VCR.

Implications and Challenges
Implications for Hardware Needs
Screen considerations

The two fundamental elements of display technology are spatial and spectral (color) resolution (Dahlberg and Jensen 1985). It can be argued that a

640×480 pixel display provides acceptable spatial resolution for most thematic maps on a standard 13-inch CRT because it decreases aliasing (jaggies) and produces an equal number of pixels per inch horizontally and vertically when the screen aspect ratio is taken into account. This resolution is available on the IBM VGA and the Macintosh II series. Higher resolutions (e.g. 800×600 or 1024×768) are also available. However, they carry higher price tags and less software has been written for them.

Although the above resolutions will produce an acceptable display, they do not compare favorably with those possible on hardcopy devices. For example, line plotters typically have resolutions approximately 10 times those of a 640×480 CRT; as a result, symbols specifically designed for the line plotter (such as Lavin and Cerveny's unit-vector method) will be difficult to display on a CRT. Another problem is that these resolutions (and screen sizes) are unacceptable when the aim is to show large maps in great detail, such as an entire U.S.G.S. topographic sheet. Although this problem can be handled in the digital world via panning and scrolling, some users may want to see the map all at once. This can only be achieved by utilizing costly high-resolution large-screen systems.

In terms of spectral resolution, both the total palette and the number of display colors must be considered. With the IBM EGA system, for example, 16 colors can be displayed at one time out of a total palette of 64. Although the number of display colors is acceptable for many applications, the total palette is not. As an illustration, selecting five evenly spaced shades of blue with EGA is virtually impossible. In contrast, IBM VGA is considerably more attractive to cartographers, because 16 colors can be selected from a palette of over 250,000.

Although being able to display 16 colors is adequate for many basic thematic maps (e.g. the classed choropleth map) it is easy to think of maps that will require a considerably greater number. Examples include unclassed choropleth maps and geological maps. Such cases are usually handled using color boards that are capable of displaying 256 colors at one time out of palettes as large as approximately 250,000 to 16.7 million. For the projected capability of multimedia systems, however, even more display colors may be needed (Miller 1989).

Speed of processing

The speed of processing required will depend on the type of map(s) to be generated. For static mapping, displays generated within a few seconds will generally be acceptable. IBM AT-compatible systems and the Apple Macintosh series computers have this capability.

For interactive and animated mapping, the speed of processing required will depend on the nature and complexity of the maps one is working with.

For two-dimensional maps (such as choropleth maps), an 80286 or 80386 processor (in the IBM-compatible world) or a Motorola 68020 or 68030 (in the Macintosh II series) would be acceptable. For three-dimensional maps one must have access to the speed of a graphics workstation environment.

Storage

Specifying the type and amount of storage space required is very difficult because it will depend on the number of spatial and attribute databases one wishes to work with. In general, the minimum requirement is a hard disk of 40 megabytes. For multimedia applications, which are likely to become increasingly popular, an optical mass storage device will be essential.

Hardcopy

Although the trend of cartography is toward interactive cartography and animated maps on display screens, this does not mean an end to the need for hardcopy. Where color is not required, current technology is more than adequate; for example, laser printers having resolutions of 300 dots per inch are available. Where color is required, as will be the case with most screen displays, current technology is costly, if accurate color rendition is desired.

Speech and music generation

Speech and music generation may at present seem superfluous to cartographers, but these are likely to be included in future map display systems. They will be integral components, for example, in multimedia systems such as those described by Miller (1989). Speech (and possibly music) may also be useful in upgrading systems that currently display maps alone. As an example, imagine an electronic atlas in which a map is shown while the computer describes the spatial pattern shown on the map. Music has less obvious possibilities, but it might be used as a way of enhancing the aesthetics of maps, as was done in the scientific visualization video (discussed above).

Implications for Cartographic Research

Based on the preceding survey, there are several potential avenues for cartographic research. One clear possibility is the need to develop new display approaches. In this respect, animation has a great deal of potential. Might Lavin and Cerveny's (1987) unit-vector density method be more effective if some form of animation were used? How might Tobler's (1987) migration mapping methods be modified through the use of animation? Particularly interesting is the potential that animation provides for portraying spatial-temporal data (Monmonier 1989b).

Another avenue for cartographic research is the need to analyze the effectiveness of the various display methods from the standpoint of the user. What is the most effective way to present spatial-temporal data through animation? Should users see all of the changes in a spatial pattern over time, or should some form of generalization be used so that major changes are more easily detected?

Real-time, three-dimensional maps and multimedia systems are particularly intriguing areas for those examining them from the standpoint of the map user. How should users manipulate such systems? Answering this question may be a complicated one for cartographers who, in previous studies, have attempted to control as many variables as possible in order to examine the nature of a particular design parameter.

A third avenue for cartographic research is the development of expert systems in relation to data display. Interactive systems often allow users to enter their own data or even design their own maps, raising the ". . . specter of poorly designed maps by the millions" (Kimerling 1989, p. 712). Expert systems may alleviate this problem by assisting the user in selecting the appropriate map and associated design parameters (see the chapter by Buttenfield and Mark).

Although expert systems are a potential area of research, their development is not without problems. One problem is determining a set of rules on which an expert system can be based. In the field of cartography, standard textbooks are available, but cartographers do not necessarily agree with the "rules" stated in those texts. Another problem is that no matter how sophisticated the expert system, some knowledge of basic cartographic principles will be required on the part of the user. As a simple example, imagine that a user wishes to map raw population data by county for the state of Kansas. Ideally, the expert system would quiz the user regarding the nature of the data, e.g. are the data counts or densities? In order to answer this question, the user would have to understand the concept of counts and densities and appreciate the ramifications of making a particular choice.

Challenges for Cartographic Education

Discussion of the limitations of expert systems makes clear the importance of a continuing focus on basic map design principles in education. Ideally, these principles should be taught within an interactive environment, rather than by traditional manual methods, so that users can focus on the principles rather than on the physical tools. Although present thematic mapping packages and CAD systems can be used in this context, specialized instructional packages would be a more suitable solution.

In addition to learning basic design principles, students will also need to learn how to manipulate and design modern data display systems. This education is, in part, closely linked with the research needs posed above. For example, research with map users is needed to determine the advantages and disadvantages of using real-time, three-dimensional maps. Teaching students how to manipulate such maps depends, in part, on the findings of such research. Students may be taught how to design modern data display systems through the use of software like HyperCard. In contrast to traditional design, where the focus is on a particular map, the focus would be on how maps link together, or how maps link with other media such as pictures, text, and speech.

Conclusion

An initial concern at the outset of this survey was the inability of digital cartography to go beyond the emulation of manual methods. After having examined a considerable number of recent references, the concern still exists, but there is no pessimism about the future. There have been interesting developments in all of the mapping categories discussed: static, interactive, and animated. In some cases the developments have been limited, or the systems required have been unavailable to most cartographers, but a review of the studies cited reveals the tremendous potential of modern data display technology.

Acknowledgements

This material is based in part upon work supported by the National Science Foundation under Grant No. SES-8706847. The authors wish to thank Professor Barbara Shortridge and Mr. Charles Ross for their helpful comments on an earlier draft.

References

Allard, L. and M. J. Hodgson (1987) "Interactive graphics for mapping location-allocation solutions", *The American Cartographer*, Vol. 14(1), pp. 49–60.

Anderson, R. H. and N. Z. Shapiro (1979) *Design Considerations for Computer-based Interactive Map Display Systems*, Report No. R-2382-ARPA, Prepared for the Defense Advanced Research Projects Agency by the Rand Corporation.

Becker, R. A. and W. S. Cleveland (1988) "Brushing scatterplots", in Cleveland, W. S. and M. E. McGill (eds.), *Dynamic Graphics for Statistics*, Wadsworth & Brooks/Cole, Pacific Grove, California, pp. 201–224.

Brassel, K. E. and J. J. Utano (1979) "Design strategies for continuous-tone area mapping", *The American Cartographer*, Vol. 6(1), pp. 39–50.

Brassel, K. E. and Z. Kiriakakis (1983) "Orthogonal three-dimensional views for thematic mapping", *Auto-Carto 6 Proceedings*, Vol. II, Ottawa, Canada, pp. 416–425.

Brown, R. F. and H. Moellering (1982) "ICFITG: A program for interactive contouring from an irregular triangular grid", *Technical Papers of the American Congress on Surveying and Mapping*, 42nd Annual Meeting, Denver, Colorado, pp. 1–15.

Calkins, H. W. (1984) "Space-time data display techniques", *Proceedings of the International Symposium on Spatial Data Handling*, Vol. II, Zurich, Switzerland, pp. 324–331.

Carstensen, L. W. (1982) "A continuous shading scheme for two-variable mapping", *Cartographica*, Vol. 19(3–4), pp. 53–70.

Carstensen, L. W. (1986a) "Bivariate choropleth mapping: the effects of axis scaling", *The American Cartographer*, Vol. 13(1), pp. 27–42.

Carstensen, L. W. (1986b) "Hypothesis testing using univariate and bivariate choropleth maps", *The American Cartographer*, Vol. 13(3), pp. 231–251.

Carter, J. R. (1984) *Computer mapping: progress in the '80s*, Association of American Geographers Resource Publication, Washington, D.C.

Clarke, K. C. (1988) "Scale-based simulation of topographic relief", *The American Cartographer*, Vol. 15(2), pp. 173–181.

Cleveland, W. S. and M. E. McGill (eds.) (1988) *Dynamic Graphics for Statistics*, Wadsworth & Brooks/Cole, Pacific Grove, California.

Cowen, D. J. (1984) "Rethinking DIDS: the next generation of interactive color mapping systems", *Cartographica*, Vol. 21(2–3), pp. 89–92.

Cowen, D. J., J. Booth, C. Heivly and P. Oppenheimer (1984) "Using color bivariate mapping procedures to model spatial processes", *Proceedings of the International Symposium on Spatial Data Handling*, Zurich, Switzerland, pp. 349–370.

Crane, G. (1988) "Redefining the book: some preliminary problems", *Academic Computing*, Vol. 2(5), pp. 6–11ff.

Cuff, D. J., J. W. Pawling and E. T. Blair (1984) "Nested value-by-area cartograms for symbolizing land use and other proportions", *Cartographica*, Vol. 21(4), pp. 1–8.

Dahlberg, R. H. and J. R. Jensen (1985) "Educational implications of the integration of cartography and remote sensing into a new conceptual model", in Taylor, D. R. F. (ed.), *Progress in Contemporary Cartography*, Vol. III, Wiley, Chichester, pp. 169–186.

Dalton, J., J. Billingsley, J. Quann and P. Bracken (1979) "Interactive color map displays of domestic information", *Computer Graphics*, Vol. 13(2), pp. 226–233.

Dent, B. D. (1985) *Principles of Thematic Map Design*, Addison Wesley, Reading, Mass.

Dobson, M. W. (1973) "Choropleth maps without class intervals? a comment", *Geographical Analysis*, Vol. 5(4), pp. 358–360.

Dobson, M. W. (1980a) "Perception of continuously shaded maps", *Annals, Association of American Geographers*, Vol. 70(1), pp. 106–107.

Dobson, M. W. (1980b) "Unclassed choropleth maps", *The American Cartographer*, Vol. 7(1), pp. 78–80.

Dobson, M. W. (1988) Personal communication.

Dudycha, D. J. (1981) "The impact of computer cartography", *Cartographica*, Vol. 18(2), pp. 116–150.

Dutton, G. H. (1979) "American graph fleeting, a computer-holograph map animation of United States population growth 1790–1970", in *Computer Mapping in Education, Research, and Medicine*, Harvard Library of Computer Graphics Mapping Collection, Laboratory for Computer Graphics and Spatial Analysis, Harvard University, pp. 53–62.

Egenhofer, M. J. and A. U. Frank (1988) "Designing object-oriented query languages for GIS: human interface aspects", *Proceedings, Third International Symposium on Spatial Data Handling*, Sydney, Australia, pp. 79–96.

Eyton, J. R. (1984) "Complementary-color, two-variable maps", *Annals, Association of American Geographers*, Vol. 74(3), pp. 477–490.

Eyton, J. R. (1986) "Digital elevation model perspective plot overlays", *Annals, Association of American Geographers*, Vol. 76(4), pp. 570–576.

Faintich, M. B. (1986) "Digital cartographic data bases: advanced analysis and display technologies", *Proceedings, Second International Symposium on Spatial Data Handling*, Seattle, Washington, pp. 600–610.

Fairchild, D. (1987) "The display of boundary information: a challenge in map design in an automated production system", *Auto-carto 8 Proceedings*, Baltimore, MD, pp. 456–465.

Gilmartin, P. P. (1987) "Tiling patterns for choropleth maps on medium resolution CRTS: an empirical solution", *Proceedings of the 13th International Cartographic Conference*, Vol. I, Morelia, Mexico, pp. 459–482.

Gilmartin, P. P. (1988) "The design of choropleth shadings for maps on 2- and 4-bit color graphics monitors", *Cartographica*, Vol. 25(4), pp. 1–10.

Goodchild, M. F. (1988) "Stepping over the line: technological constraints and the new cartography", *The American Cartographer*, Vol. 15(3), pp. 311–319.

Goulette, A. M. (1985) "Cartographic use of dithered patterns on 8-color computer monitors", *Auto-Carto 7 Proceedings*, Washington, D.C., pp. 205–209.

Grogan, D. L. (1986) "Perspective raster mapping", Unpublished M.S. thesis, Department of Geography, The Pennsylvania State University.

Groop, R. and J. Harman (1988) "Cartographic representation of transitional boundaries", Abstract, Association of American Geographers Annual Meeting, Phoenix, Arizona, p. 68.

Groop, R. E. and P. Smith (1982a) "A dot matrix method of portraying continuous statistical surfaces", *The American Cartographer*, Vol. 9(2), pp. 123–130.

Groop, R. E. and R. M. Smith (1982b) "Matrix line printer maps", *The American Cartographer*, Vol. 9(1), pp. 19–24.

Hartnett, S. (1987) "Employing rectangular point symbols in two-variable maps", Abstract, Association of American Geographers Annual Meeting, Portland, Oregon, p. 39.

Jenks, G. F. (1989) Personal communication.

Junkin, B. G. (1982) "Development of three-dimensional spatial displays using a geographically based information system", *Photogrammetric Engineering and Remote Sensing*, Vol. 48(4), pp. 577–586.

Kimerling, A. J. (1989) "Cartography", in Gaile, G. L. and C. J. Willmott (eds.), *Geography in America*, Merrill Publishing Company, Columbus, Ohio, 686–718.

Kimerling, A. J. and H. Moellering (1989) "The development of digital slope-aspect displays", *Auto-Carto 9 Proceedings*, Baltimore, MD, pp. 241–244.

Lai, P. (1985) "Moving towards an interactive tactual mapping environment", *The Cartographic Journal*, Vol. 22(2), pp. 102–105.

Lam, N. S. (1983) "Spatial interpolation methods: a review", *The American Cartographer*, Vol. 10(2), pp. 129–149.

Lang, L. (1989) "GIS goes 3D: as the GIS sheds its 2D mold, exciting 3D applications begin to emerge", *Computer Graphics World*, Vol. 12(3), pp. 38–46.

Lavin, S. (1986) "Mapping continuous geographical distributions using dot-density shading", *The American Cartographer*, Vol. 13(2), pp. 140–150.

Lavin, S. and J. C. Archer (1984) "Computer-produced unclassed bivariate choropleth maps", *The American Cartographer*, Vol. 11(1), pp. 49–57.

Lavin, S. J. and R. S. Cerveny (1987) "Unit-vector density mapping", *The Cartographic Journal*, Vol. 24(2), pp. 131–141.

Lippincott, R. (1990) "Beyond hype", *Byte*, Vol. 15(1), pp. 215–218.

Marble, D. F. (1987) "Introduction" to articles dealing with the computer and cartography, *The American Cartographer*, Vol. 14(2), pp. 101–103.

McCormick, B. H., T. A. DeFanti and M. D. Brown (1987) "Visualization in scientific computing", *Computer Graphics*, 21(6).

McGranaghan, M., D. M. Mark and M. D. Gould (1987) "Automated provision of navigation assistance to drivers", *The American Cartographer*, Vol. 14(2), pp. 121–138.

McLaren, R. A. (1989) "Visualization techniques and applications within GIS", *Auto-Carto 9 Proceedings*, Baltimore, MD, pp. 5–14.

Meyer, M. A., F. R. Broome and R. H. Schweitzer Jr. (1975) "Color statistical mapping by the U.S. Bureau of the Census", *The American Cartographer*, Vol. 2(2), pp. 100–117.

Miller, D. W. (1988) "The great American history machine", *Academic Computing*, Vol. 3(3), pp. 28–29ff.

Miller, M. J. (1989) "Multimedia technology is not just the buzzword of the year", *InfoWorld*, Vol. 11(15), pp. 56–57.

Moellering, H. (1978) *A Demonstration of the Real-Time Display of Three-Dimensional Cartographic Objects*, Computer animated videotape, Department of Geography, Ohio State University, Columbus.

Moellering, H. (1980) "The real-time animation of three-dimensional maps", *The American Cartographer*, Vol. 7(1), pp. 67–75.

Moellering, H. (1984) "Real maps, virtual maps and interactive cartography", in Gaile, G. L. and C. J. Willmott (eds), *Spatial Statistics and Models*, D. Reidel Publishing Co., Dordrecht, The Netherlands, pp. 109–132.

Moellering, H. (1989a) "A practical and efficient approach to the stereoscopic display and manipulation of cartographic objects", *Auto-Carto 9 Proceedings*, Baltimore, MD, pp. 1–4.

Moellering, H. (1989b) "An analytical approach to the cartographic display of satellite data in two and three dimensions", Paper to be published for the International Cartographic Association meetings, Budapest, Hungary.

Monmonier, M. S. (1978) "Modifications of the choropleth technique to communicate correlation", *International Yearbook of Cartography*, Vol. 18, pp. 143–157.

Monmonier, M. S. (1982a) "Cartography, geographic information, and public policy", *Journal of Geography in Higher Eduction*, Vol. 6(2), pp. 99–107.

Monmonier, M. S. (1982b) *Computer-Assisted Cartography : Principles and Prospects*, Prentice-Hall, Englewood Cliffs, New Jersey.

Monmonier, M. (1989a) "Geographic brushing: enhancing exploratory analysis of the scatterplot matrix", *Geographical Analysis*, Vol. 21(1), pp. 81–84.

Monmonier, M. (1989b) "Visualization of geographic time-series data: a conceptual framework", Abstract, Association of American Geographers Annual Meeting, Baltimore, Maryland, p. 145.

Morrison, J. L. (1986) "Cartography: a milestone and its future", *Proceedings, Auto Carto, London*, Vol. 1, pp. 1–12.

Muller, J.-C. (1980) "Comment in reply", *Annals, Association of American Geographers*, Vol. 70 (1), pp. 107–108.

Muller, J.-C. (1989) "Challenges ahead for the mapping profession", *Auto-Carto 9 Proceedings*, Baltimore, MD, pp. 675–683.

Nickerson, B. G. (1988) "Automated cartographic generalization for linear features", *Cartographica*, Vol. 25(3), pp. 15–66.

O'Callaghan, J. F. and Simons, L. W. (1984) "Map display techniques for interactive colour mapping", *Proceedings of the International Symposium on Spatial Data Handling*, Vol. II, Zurich, Switzerland, pp. 316–323.

Olson, J. M. (1976) "Non-contiguous area cartograms", *The Professional Geographer*, Vol. 28(4), 371–380.

Olson, J. M., C. A. Brewer, J. B. Moore and M. H. Martel (1989) "Maps for the color deficient: experimental results", Abstract, Association of American Geographers Annual Meeting, Baltimore, Maryland, p. 157.

Openshaw, S. and H. Mounsey (1987) "Geographic information systems and the BBC's Domesday interactive videodisk", *International Journal of Geographical Information Systems*, Vol. 1(2), pp. 173–179.

Plumb, G. A. and T. A. Slocum (1986) "Alternative designs for dot-matrix printer maps", *The American Cartographer*, Vol. 13(2), pp. 121–133.

Poiker, T. K. and L. A. Griswold (1985) "A step toward interactive displays of digital elevation models", *Auto-carto 7 Proceedings*, Washington, D.C., pp. 408–415.

Robb, M. C. (1987) "CARTRIPS – a cartographic route information presentation system", *The Cartographic Journal*, Vol. 24(1), pp. 42–49.

Robertson, P. K. (1988) "Choosing data representations for the effective visualization of spatial data", *Proceedings, Third International Symposium on Spatial Data Handling*, Sydney, Australia, pp. 243–252.

Robertson, P. K. and J. F. O'Callaghan (1985) "The application of scene synthesis techniques to the display of multi-dimensional image data", *ACM Transactions on Graphics*, Vol. 4(4), pp. 247–275.

Robinson, P. (1990) "The four multimedia gospels", *Byte*, Vol. 15(1), pp. 203–212.

Sibert, J. L. (1980) "Continuous-color choropleth maps", *Geo-Processing*, Vol. 1, pp. 207–216.

Siekierska, E. (1984) "Towards an electronic atlas", *Cartographica*, Vol. 21(2–3), pp. 110–120.

Siekierska, E. M. and S. Palko (1986) "Canada's electronic atlas", *Proceedings, Auto Carto London*, Vol. II, pp. 409–417.

Slocum, T. A., S. L. Egbert, M. C. Prante and S. H. Robeson (1988) "Developing an information system for choropleth maps", *Proceedings, Third International Symposium on Spatial Data Handling*, Sydney, Australia, pp. 293–305.

Smith, D. R. and A. R. Paradis (1989) "Three-dimensional GIS for the earth sciences", *Auto-Carto 9 Proceedings*, Baltimore, MD, pp. 324–335.

Smith, R. M. (1977) "The development of black and white two-variable maps: ongoing cartographic research", in Winters, H. A. and M. K. Winters (eds.), *Applications of Geographic Research*, Department of Geography, Michigan State University, East Lansing, pp. 145–154.

Smith, R. M. (1987) "Electronic atlas of Arkansas: design and operational considerations", *Proceedings of the 13th International Cartographic Conference*, Vol. IV, Morelia, Mexico, pp. 161–167.

Steiner, D. R., M. J. Egenhofer and A. U. Frank (1989) "An object-oriented carto-graphic output package", *Technical Papers, ASPRS/ACSM Annual Convention*, 5 (Surveying and Cartography), Baltimore, MD, pp. 104–113.

Taylor, D. R. F. (1984) "The creation and design of maps for videotex systems", *Technical Papers of the Austra Carto One Seminar*, Perth, Australia, pp. 277–289.

Taylor, D. R. F. (1985) "The educational challenges of a new cartography", *Cartographica*, Vol. 22(4), pp. 19–37.

Taylor, D. R. F. (1987) "Cartographic communication on computer screens: the effect of sequential presentation of map information", *Proceedings of the 13th International Cartographic Conference*, Vol. I, Morelia, Mexico, pp. 591–611.

Tobler, W. R. (1965) "Automation in the preparation of thematic maps", *The Cartographic Journal*, Vol. 2(1), pp. 32–38.

Tobler, W. R. (1973) "Choropleth maps without class intervals", *Geographical Analysis*, Vol. 5(3), pp. 262–265.

Tobler, W. R. (1987) "Experiments in migration mapping by computer", *The American Cartographer*, Vol. 14(2), pp. 155–163.

Tomlinson, R. F. (1988) "The impact of the transition from analogue to digital cartographic representation", *The American Cartographer*, Vol. 15(3), pp. 249–261.

Turner, S. H. and Moellering (1979) "ICMS: An interactive choropleth mapping system", *Proceedings of the American Congress on Surveying and Mapping*, 39th Annual Meeting, Washington, D.C., pp. 255–268.

van Elzakker, C. P. J. M. and F. J. Ormeling (1984) "Computer-assisted statistical mapping systems: user requirements", *Technical Papers of the 12th Conference of the International Cartographic Association*, Vol. I, Perth, Australia, pp. 535–551.

Vaughan, T. (1988) *Using Hypercard: from Home to Hypertalk*, Que Corporation, Carmel, Indiana.

Wainer, H. and D. Thissen (1981) "Graphical data analysis", *Annual Review of Psychology*, Vol. 32, pp. 191–241.

Waters, N. M. and G. J. A. De Leeuw (1987) "Computer atlases to complement printed atlases", *Cartographica*, Vol. 24(1), pp. 118–133.

Weissman, R. F. (1988) "From the personal computer to the scholar's workstation", *Academic Computing*, Vol. 3(3), pp. 10–14ff.

White, D. (1985) "Relief modulated thematic mapping by computer", *The American Cartographer*, Vol. 12(1), pp. 62–67.

Yoeli, P. (1983) "Shadowed contours with computer and plotter", *The American Cartographer*, Vol. 10(2), pp. 101–110.

Yoeli, P. (1984) "Cartographic contouring with computer and plotter", *The American Cartographer*, Vol. 11(2), pp. 139–155.

Yoeli, P. (1985a) "The making of intervisibility maps with computer and plotter", *Cartographica*, Vol. 22(3), pp. 88–103.

Yoeli, P. (1985b) "Topographical relief depiction by hachures with computer and plotter", *The Cartographic Journal*, Vol. 22(2), pp. 111–124.

Yoeli, P. (1986) "Computer executed production of a regular grid of height points from digital contours", *The American Cartographer*, Vol. 13(3), pp. 219–229.

CHAPTER 10

Micro-Atlases and the Diffusion of Geographic Information: An Experiment with HyperCard

JEAN-L. RAVENEAU, MARC MILLER, YVES BROUSSEAU and
CLAUDE DUFOUR

Department of Geography
Laval University Quebec
Quebec, Canada

Introduction

With the multiplication of geographic information systems (GIS) in government and the private sector, a large amount of spatially referenced data is potentially available for diffusion to the public. But these data remain underused due to a lack of processing and the absence of efficient means of communication. There is a need for the transformation of spatial data contained in computer data banks into significant spatial information, and for the diffusion of that information by means of attractive and efficient visual displays (see chapter by Slocum and Egbert). The electronic atlas is one of the possible solutions.

The purpose of this chapter is to show how it is possible to communicate more efficiently the information contained in a spatially referenced data bank by using micro-atlases designed on a microcomputer and exploiting the hypertext concept. We will examine the main aspects of the conception and creation of two electronic atlases developed on a Macintosh computer with the HyperCard software. These micro-atlases contain maps, graphs, diagrams, text, pictures and even sounds that can be accessed by means of a navigation structure.

After examining the general context for the development of electronic atlases and the potential of the hypertext concept for the structuration and visualization of geographic information, the main characteristics of two micro-atlases, one devoted to North American French communities and the other to the geography of mines and minerals in Canada will be analyzed.

The General Context for the Development of Electronic Atlases

Recent years have witnessed the development, at an accelerated pace, of new technologies and research methods that offer great potential for application in the areas of management, data processing, graphic representation and communication of geographic data and information. Several factors have contributed to this situation:

(1) The phenomenal development of microcomputing has made available to individuals and institutions microcomputers that are more and more powerful and performing at a continuously lower cost.

(2) Geographic data bases have been created or are in the process of being developed at several places; they may contain either digital base maps (i.e. digital maps from Energy, Mines and Resources Canada) or spatially referenced thematic data (i.e. geographically referenced data from Statistics Canada). Full geographic information systems, whose data can be used by specialists or by the general public are now available in Canada. The Canada Lands Data Systems (CLDS, Environment Canada) and the National Atlas Information System (NAIS) are good examples of these systems.

(3) Recent research on the processing of geographic information for instructional purposes has contributed to a renewal in the field of conception and production of instructional atlases as well as the communication of geographic information in general (see Marcotte and Tessier 1987; Raveneau *et al.* 1987; Carswell *et al.* 1987). The publication of atlases such as the *Junior Atlas of Alberta* (Alberta Education 1979) and *L'interAtlas* (Centre d'études en enseignement du Canada 1986), prepared according to these new standards mark a turning point in methodological research on school atlases in Canada.

(4) The integration of geographic information within computerized information systems has contributed to the concept of "electronic atlas". A prototype of this new kind of atlas is being developed at the Department of Energy, Mines and Resources in Ottawa (Siekierska and Palko 1986). Several similar experiments have produced operational electronic atlases. One of the most ambitious is the *Domesday Project*, prepared by BBC Enterprises (Openshaw and Mounsey 1986); another, more modest, is the *Electronic Atlas of Arkansas* (Smith 1987).

(5) The diffusion of the hypertext concept and its translation into such software as HyperCard facilitates the application of structured geographic information to an electronic atlas.

All these conditions taken together constitute a favourable context for finding new ways to process and display geographic information. It is not sufficient to use the computer to reproduce on a screen what was formerly represented on conventional maps or in traditional atlases, but rather to make use of its potential for developing new methods facilitating research, communication and the acquisition of knowledge in the field of geographic information as contained in spatially referenced information systems, as several authors in this volume have argued.

Potential of the Hypertext Concept for the Structuration and Visualization of Geographic Information

Visualization of Structured Models Depicting Geographical Knowledge

Traditionally, printed atlases contained a series of maps ordered according to a more or less definite plan. The past decade has seen the multiplication of atlases that no longer present only maps, but also text, photographs, diagrams and graphs in order to illustrate the various characteristics of geographic phenomena (Tessier 1989). The best of these atlases seek to demonstrate interrelations between phenomena so as to facilitate the understanding and explanation of the organization of geographic space. Such characteristics are exemplified in *L'interAtlas* in which the images depicting geographical information are organized along a pre-determined reading path designed according to an established model (i.e. the systemic model) to reflect the logical organization of each subject illustrated. Despite its interest, this approach is limited by the number of images that can be placed on a printed double page.

In a computer environment the limitations concerning the number of images that one can process and display are determined by the memory available. However, in order to present structured geographic information, it is necessary to be able to retrieve and link images in multiple directions along different reading paths. Nevertheless, one must be able to maintain a link between what is on the screen and the general model of organization pertaining to the subject presented (semantic model). It is also necessary to be able to easily establish relations between the different elements comprising geographical knowledge by using information contained in the data base. Hypertext is a means for operationalizing this step.

The Hypertext Concept

Hypertext is not a new concept. It was formulated for the first time by V. Bush in a paper published in 1945 (Bush 1945). The word "hypertext" refers to a non sequential disposition of information (specifically written information). This disposition is intended to imitate a mental process: "The structure of ideas is never sequential; and indeed, our thought processes are not very sequential either. True, only a few thoughts at a time pass across the central screen of the mind; but as you consider a thing, your thoughts crisscross it constantly . . ." (Nelson 1967).

The technology existing in 1945 did not allow exploitation of the power of this concept. It was necessary to await the work of Englebart and Nelson (Englebart 1963; Nelson 1967) and the advent of powerful computers and sophisticated interfaces to recognize the full potential of this concept. Amid the myriad uses of hypertext, that of browsing systems offers many benefits for teaching, public use and reference systems. Indeed, the power of this approach results from its ability to establish swiftly semantic links between knowledge emanating from diverse media. The ZOG system, developed since 1972 at Carnegie Mellon University, is organized around the concept of "frames"; it was introduced in 1982 into an information system aboard a U.S. Navy ship (USS *Carl Vinson*). The Shneiderman research team from the University of Maryland developed the "hyperties" system in order to demonstrate the possibilities of this approach for individual acquisition of thematic knowledge (Shneiderman and Morariu 1986). This system has been implemented recently in a Washington museum to aid visitors "navigate" among the thematic elements depicting the Jewish holocaust in Austria. In this application, the visitor is guided along a path while at the same time having the opportunity to refer to a variety of information by making maximum use of semantic references. These references allow the visitor to delve momentarily into a particular idea without breaking from the main thematic path.

The availability of HyperCard software, released for the first time in 1987 by the Apple Corporation, has made possible the development of user friendly systems easily available to a greater number of personal computer users (see Cartier 1988; Goodman 1987; Shafer 1988; and the chapter by Keller and Waters in this volume). Two factors have contributed to the success of this software:

(1) The ease with which one can develop stacks (HyperCard documents) and establish links between their elements, by using an advanced language (HyperTalk) of the OOP (Object Oriented Programming) type; HyperTalk allows easy and original translating of the hypertext concept to a specific structure of information.

(2) The diffusion of HyperCard stacks is facilitated by the ready availability of the software which is distributed free of charge to every new buyer of a Macintosh computer.

The present technology of icons and windows enables, through the exploitation of the visual power of images which assume different shapes, the creation of hypertext system databases that need heterogeneous representation structures. Its application to the learning and communication of geographic knowledge to a wide range of audiences appears to be very promising. However, the potential of the hypertext concept for the transmission of geographic knowledge is not tied exclusively to the HyperCard software. Other softwares, running on different types of equipment, make use of the hypertext concept (Caron 1989). Some are distributed on a commercial basis: Guide, running on IBM PC and Macintoshes and KMS, running on Sun and Apollo stations. Others serve mainly as research tools (Neptune) or are in the process of development (Xanadu). But none of these has as yet obtained the same level of diffusion and acceptance as HyperCard.

The Potential of Hypertext and HyperCard for the Communication of Geographic Information

The ability of HyperCard software to make the hypertext concept operational is accompanied by two very interesting properties useful for the communication of geographic information:

— the possibility of making multiple associations between different elements of information;
— the possibility of representing and visualizing information by various graphic means.

The possibility to make multiple associations between different elements of information is an advantage for illustrating and demonstrating interrelations that exist between geographic phenomena or between different aspects of the same phenomenon. These interrelations can be predefined for didactic purposes and translated into a navigation structure useful to the user in search of information. The hypertext concept is extremely democratic, granting maximum freedom of choice to the user to be master of this process. These possibilities become even more interesting with a software like HyperCard which allows the forging of links between a variety of media sources. It is possible to combine text, maps, graphs, diagrams, photographs, video sequences, etc., all to the accompaniment of sound and visual effects. The perception of geographic phenomena at different scales can also be illustrated in dramatic fashion with successive scale enlargements or reductions. With HyperCard, the hypertext concept is transformed into hypermedia, opening the way to numerous applications, such as the multi-media electronic atlas. It is in this context that a research group working in the Department of Geography at l'Université Laval (Québec City, Canada) has developed two micro electronic atlases:

— La Francophonie nord-américaine à la carte (North American French-speaking communities à la carte);
— Mines et minéraux à la carte (Mines and Minerals à la carte).

Two Examples of Electronic Atlases Produced Using HyperCard

The two micro-atlases, *La Francophonie nord-américaine à la carte* and *Mines et minéraux à la carte*, have several features in common;

— they have been constructed using data from a spatially referenced data base;
— they have been developed using the HyperCard software and run on Macintosh microcomputers;
— they present geographic information in a multi-media display;
— they present geographic information in a structured form, but allow also free undirected exploration of this information.

However, when the two micro-atlases are examined individually, differences in the method used for their elaboration and in the information structure are observable.

The Micro-Atlas de la Francophonie en Amérique du Nord

The idea for this electronic atlas (Brousseau and Miller 1988) was born when the time came to update the cartographic work of Louder, Morissonneau and Waddell (1979). It was only fitting to use the new means at hand. The two main goals were:

— Use computer technology to process the geographic information.
— Propose new ways for visualizing the geographic data that would be both more accessible to the layman and more visually attractive.

How these goals were achieved is the subject of a M.A. thesis written by Y. Brousseau (Brousseau 1988).

A Spatial Database

Canadian (1981) and US (1980) census data were already available on tapes on the basis of 3100 U.S. counties, 1340 Canadian census divisions and 36 metropolitan areas. At that scale, it was judged inappropriate to retain county and divisional boundaries. Thus only the centroids were noted for the purpose of locating the spatial units. The pre-digitized boundaries of the Canadian provinces and American states were imported from a public file (SAS/GRAPH).

The First Electronic Atlas: A Series of Maps

Making use of Thema (Nolette 1986), a software package developed and used at the Laval University Geography Department and the thematic and geographic information already available, a map of the distribution of the

French communities in North America was produced. Although interesting, this initial step proved to be somewhat limited in its degree of usefulness. A series of maps at a larger "regional" scale was then constructed and brought back from the mainframe to a microcomputer of the IBM PC family. Using two software packages, one to draw the different maps on the microcomputer screen (PcPlot) and the other to automate the selection process of the maps (SmartKey), it was possible to create a first electronic atlas. Aside from its "modern" flavour, it added little in the way of analysing and interpreting the geographic information. Further, owing to legal considerations, it was impossible to distribute the atlas on anything but a very small scale.

From Static to Dynamic

When Apple released HyperCard in 1987 it was seen as a new tool with tremendous potential for the development of electronic atlases on micro-computers. This new medium offered a way to create, from the already available data, an atlas which would be both more visually oriented and more user friendly than the previous PC version. The fact that HyperCard was to be included with every Macintosh sold would also solve the distribution problem. It was thus decided to use HyperCard and the Macintosh to make a new and more complete version of the atlas. To do so, it was first necessary to transfer the work already done on the mainframe computer to a Macintosh environment.

From Mainframe to Micro

Using appropriate software, it was a simple task to transfer the maps from the mainframe to a Macintosh PICT format, but since the screen format of the original Tektronik image (1024×768) was not the same as the Mac (512×342), it was necessary to further edit the resulting images via Canvas. The corrected maps were then transferred to HyperCard using the clipboard. These transformations are illustrated in Fig. 10.1.

The Content

The information presented in the atlas consists of three statistical variables and eleven geographic regions. The three variables are: Total population of French origin by sector (the "origine" button in the atlas); total population speaking French at home (the "parlé" button) and, finally, a retention index (the "rétention" button) calculated using the two other variables. The eleven regions are at different scales (Figs. 10.2 and 10.3): North America [Canada (Western Provinces, Ontario, Quebec, Atlantic Provinces), United States (New England, Southeast, Southwest, Midwest)]. The use of proportional circles which maintain a constant scale from map to map helps the user wishing to make comparisons (Figs. 10.3 and 10.4). Furthermore, additional

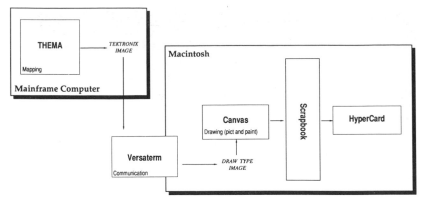

FIG. 10.1. From mainframe to micro: the software path.

information of a cultural nature consisting of textual and visual descriptions of the flag of each French community have been included where relevant (Fig. 10.5).

A Dynamic Structure

To make the transition between a traditional series of maps and a dynamic electronic atlas where the user can browse freely but coherently, it was necessary to create a navigation structure. Once the user has access to the micro-atlas, he or she can click on one of ten navigation buttons (Fig. 10.6):

Instructions	To access the "on line" help card
Impression	To print the current card
Stop	To leave the atlas or return to the beginning
Région	To change region
Localisation	To see a location map for the current region
Origine, Parlé, Rétention	To see the maps pertaining to the three thematic variables
Small question mark	To see a brief description of the variables
Flag	To see the flag of the local French community

The user may choose to explore a single thematic variable passing from region to region, or to look at the different variables within the same region. It is also possible to consult the micro-atlas in a completely free or randomized way, switching from one region to another or from one theme to another in any order, and going back to thematic variables information or

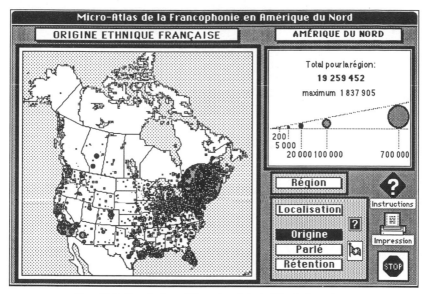

FIG. 10.2. Population of French origin in North America.

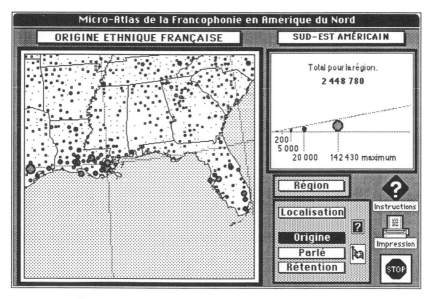

FIG. 10.3. Population of French origin in Southeast U.S.

contextual help whenever necessary. The navigation buttons are always available on the screen; any push on the appropriate button will bring the selected option.

To go from one image to another is not a haphazard process: five different

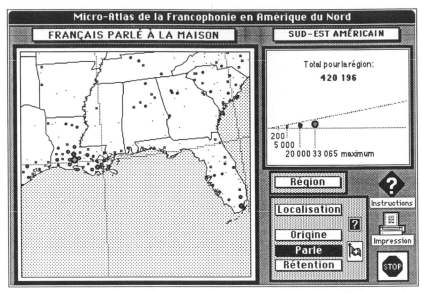

FIG. 10.4. Population speaking French at home in Southeast U.S.

FIG. 10.5. Flag of one of the French communities of North America (Louisiana).

visual effects (dissolve, checkerboard, scroll, wipe, barndoor) have been used in order to remind the user of the transition between the different parts of the atlas. For example, a slow dissolve is used to switch from the location map to another variable within the same region. This same effect has also

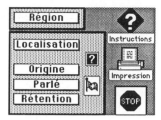

FIG. 10.6. Navigation buttons controlling the access to the different parts of the atlas.

been used to switch from home spoken language map to French ethnic origin map. This effect is powerful and allows easy comparison of two variables, as proportional circles use the same scale. It is argued that there is no printed representation that could result in as vivid a comparison of maps. If the user requests access to another region, the scroll down effect simulates a page gliding over another one. Another very interesting effect occurs if the user wishes to know who is responsible for the production of the atlas: the screen splits open and will close upon request, as if a book were opened and then closed as soon as the information had been grasped.

Sound effects start with the opening of the atlas in order to draw the users attention and give the atlas an impressive professional touch. These sound and visual effects bring a positive psychological influence upon the user; it is a simple technique although quite effective. It helps the user to move around the atlas effortlessly within a pleasant environment. To consult the atlas is more of a game than a task to be executed.

Mines and Minerals à la Carte

La Francophonie nord-américaine à la carte constitutes a first attempt at the graphic representation and dynamic management of a collection of electronic maps using HyperCard. However, the content and information structure of this atlas remains relatively simple. In a second application, concerning the geography of mines and minerals in Canada, the potential of HyperCard was tested in the visual presentation of more complex geographic information (Raveneau, Dufour and Miller 1989). The goal was to verify the software's ability to communicate visually interrelations between several components of the same phenomena by means of maps, graphs, diagrams, drawings, text, etc., illustrating spatially referenced statistical data.

State of Data on the Geography of Mines and Minerals: In Search of Suitable Data

There exists an abundance of data on the geography of mines and minerals collected by public services such as the Energy, Mines and Resources

Canada, Energy and Resources Québec, as well as by private organizations such as the Mining Association of Canada. These organizations publish yearbooks or statistical repertories and maps containing detailed data spatially referenced at different scales (see Canada 1988; Québec 1988; Canadian Mining Association 1988). This documentation can be obtained fairly easily in printed form and is well adapted to fulfil detailed information needs on a particular activity or a specific region. However, this documentation remains underused to a large extent for two reasons: (1) its dispersion and the public ignorance of its existence; (2) data processing not conceived for use by the general public. The elements of a spatially referenced data bank on mines and minerals does indeed exist, but they have not yet been organized in an operational manner for a large diffusion by electronic means. Moreover, spatially referenced statistical data frequently exist only in the form of unprocessed tables and are thus difficult to use for analysis of a particular activity or region. Therein lies an important lacuna that prevents the full use of available data, especially by a non-specialized public composed of technocrats, teachers, students and ordinary citizens. Two solutions can be proposed to alleviate this problem:

(1) Establish a geographic information system on mines and minerals.
(2) Design new ways for communicating geographic information so that it might be more readily visualized and interpreted by the general public; one of these new ways is the electronic atlas.

The second solution is obviously complementary to the first since the availability of data already coded in computer format facilitates the conception of an electronic atlas and the development of new forms of visualization of geographic information. Inasmuch as a geographical information system on mines and minerals did not exist, it was necessary to create a data base. The data were extracted from printed tables in numerous publications (*Canadian Mineral Yearbook*, Canadian Mining Association, etc.).

Organizing Information: A Systemic Perspective

It was decided to illustrate, for educational purposes, the information pertaining to the geography of mines and minerals in Canada, for the secondary schools public. Curriculum objectives of the geography program at this level were selected focusing upon interrelations between components of the topic. The experience gained from the preparation of *L'interAtlas* (Centre d'Études en Enseignement du Canada 1986; see plates 32 to 37) proved invaluable. The seventeen most important minerals or metals in Canada were selected and classified according to their production value (fuels excluded). Data pertaining to each mineral or metal are assigned to one of eleven subjects (Table 10.1). The ordering of the subjects reflects a logical sequence based on a systemic approach. The different aspects of mineral production are presented in an integrated manner: physical description and use of each mineral, extraction and transformation processes, location of mines at different scales (Canada, Québec, regions), mining landscapes, importance of production at different scales (World, Canada, Provinces, Québec) and evolution of this production over a 10-year period (1978–87), international trade (exportations and importations); other information, concerning manpower, environmental issues, etc., have not been documented in the present version of the micro-atlas. Data on the geography of mines and minerals are organized in the form of an information matrix composed of 12 rows and 17 columns (Fig. 10.7) which may be

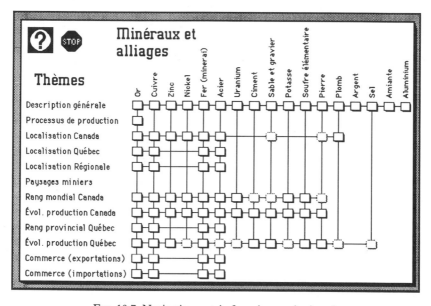

FIG. 10.7. Navigation matrix for mines and minerals.

addressed by row, by column, by individual cell or by group of cells. This matrix served as a conceptual framework for data input and for the construction of a navigation structure in HyperCard.

Data Input and Stack Construction with HyperCard

Each subject (row) in the matrix corresponds to a set of homogeneous data that are presented visually in different forms (see Table 10.1): Text, graphs, diagrams, drawings, maps, photographs (Fig. 10.8, a to f); sound could

TABLE 10.1

List and Presentation Forms of Subjects in the Micro-Atlas "Mines et minéraux à la carte" (Mines and Minerals à la carte)

Subjects	Type of document
General description: properties, uses, substitutes, recycling	Text, graphs
Production processes: extraction methods, primary transformation, transportation	Diagrams, text
Location of mines: in Canada, in Québec and by region	Maps, text
* Mining landscapes	Scanned photographs, diagrams
* Environmental problems caused by mining activities	Text, scanned photographs, diagrams
Production statistics for Canada (economic aspects):	
● world rank of Canada	Maps
● evolution of Canadian production, by tonnage and value (1978–1988)	Graphs
● evolution of the percentage of each mineral in the total production	Graphs
● production by province	Maps, graphs
Production statistics for the Province of Québec:	
● evolution of Québec production (1978–1988)	Graphs
● evolution of the percentage of each mineral in the total production	Graphs
International trade: exports, Canada	
● main countries of destination for exports (in value)	Maps
● evolution of exports	Graphs
International trade: imports, Canada	
● main countries of origin for imports (in value)	Maps
● evolution of imports	Graphs
* Labour force:	
● distribution by province	Maps
● evolution	Graphs
* General summary: international conjuncture, financial investments, mining policies and their impact, etc.	Text

*Note: The topics marked by an * are not yet documented in the present version (1989) of the micro-atlas.*

have been added if desired. In HyperCard format, each matrix row is materialized by a stack of cards, each one containing a piece of information represented in one or several of the forms cited above.

Most of the individual data entered in each stack were processed in different ways following their extraction from original printed sources comprised of text, statistical tables or maps. All the images included (maps, graphs, diagrams) are original creations, constructed from raw data by using

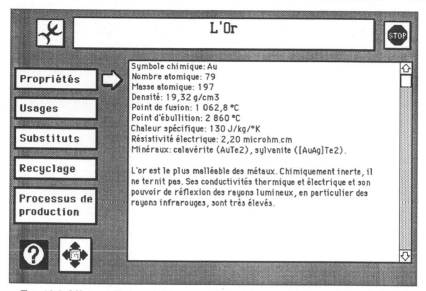

FIG. 10.8. Mines and minerals à la carte: selected images. (a) General description of gold. (b) Production process of gold. (c) Location of gold mines in Canada. (d) Description of copper producers in Timmins, Ontario. (e) Zoom on the Abitibi region. (f) Evolution in the production of gold in Canada.

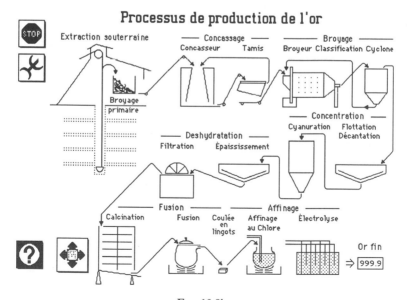

FIG. 10.8b.

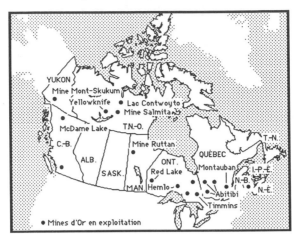

FIG. 10.8c.

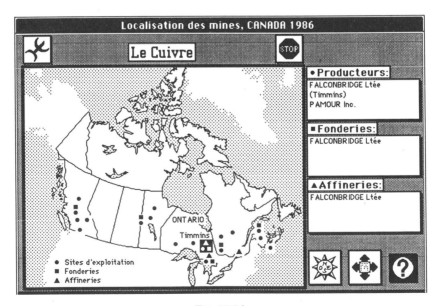

FIG. 10.8d.

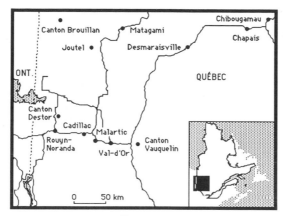

FIG. 10.8e.

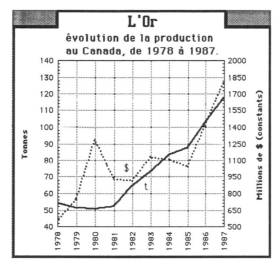

FIG. 10.8f.

several graphic or cartographic software package in a Macintosh or MS-Dos environment (Fig. 10.9). After their creation these images were imported to HyperCard and completed with titles and text; navigation buttons were added in order to allow the navigation from mineral to mineral or from subject to subject (Fig. 10.10).

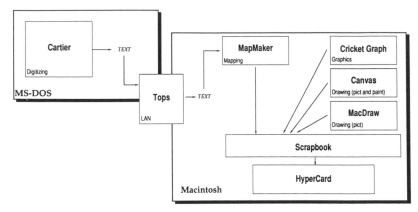

FIG. 10.9. Mines and minerals: the software.

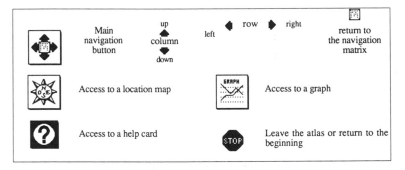

FIG. 10.10. Mines and minerals: the navigation buttons.

Navigation Structure and the Different Ways of Discovering Information

The matricial ordering of information is a well adapted structure for the presentation of the geography of mines and minerals in Canada. This structure was selected as a model for designing the navigation board for the atlas (Fig. 10.7). Two paths are possible:

— a vertical path (column) for an integrated study of a set of geographic variables concerning one mineral;

— a horizontal path (row) for a comparative study of one topic through several minerals.

These two ways for discovering information are semi-directed; they are based on a logical approach to the geography of mines and minerals as elaborated from instructional objectives. But the user remains free to start from any cell in the matrix and to visualize information following a completely random path. The direction of navigation is chosen by clicking on a navigation button which offers five directions: up, down, right, left and return to the main matrix (navigation board). This freedom of movement is made possible by an appropriate programming in HyperTalk language which, in turn, enables maximum use of the hypertext concept.

Some Important Facts Concerning the Programming of the Navigation Structure

With HyperCard it is possible to label and program each object: a button, a field, a card, a background or a stack. By using significant labels and by placing the right program in the right object, each one being at a different hierarchical level, it is possible to minimize programming and optimize the updating. For example, in the navigation stack (Fig. 10.7), each button of the matrix gives the user an access to information pertaining to a particular topic or mineral. It would have been possible to program these buttons as follows: "When the user clicks on this button, go to the such-and-such a card pertaining to the such-and-such a stack." Each button would then had to have a different program. Instead, each button was labelled according to its location in the matrix: For example, the button corresponding to the location of mines in Canada bears the name CA (row C and column A of the matrix). In the same way, each row corresponds to a stack, whose name is defined by its location into the matrix (A to L). The cards within each stack are named according to the same rule. Making use of this structure the programming instructions at the stack level may resemble this: "Check which button was activated by the user and go to the card whose name is identical (to that of the button) and which is situated in the stack whose name corresponds to the first letter of the button name." This kind of programming may seem more complex but the result is one program that controls 204 buttons instead of 204 separate programs. More programming tips could be described, but suffice it to say that a wise use of the HyperTalk language is a convenient means to exploit, in an efficient manner, the full potential of the hypertext concept. This object oriented programming renders the programming on Macintosh microcomputers accessible to many people and is especially well adapted to the design of micro-atlases.

A Summary of the Experience of Creating Two Micro-Atlases with HyperCard

Positive Aspects

The most attractive aspect of an electronic atlas constructed using HyperCard is the immense didactic potential of the hypertext concept which enables the user to "navigate" through geographic information and to establish multiple relationships between data illustrated by various type of visual representation. The user is no longer limited by the physical format of a printed page to establish relations between several images that can be called and viewed at will. However, this didactic potential will be more evident for the user if the information presented in the atlas has been organized beforehand, according to a logical structure. In the absence of a logical predefined path for orienting the user in the atlas, he or she may rapidly get lost in the intricacy of information that are presented, especially if he or she has no particular idea of what to visualize.

Finally, HyperCard has the great advantage of permitting the creation of electronic atlases at low cost; the minimal hardware and software require less than $5,000.

Limitations

An important limitation is imposed by the small display format of HyperCard (version 1.2) on standard Macintosh screens measuring 23 cm with a definition of 512×342 pixels. The small dimension of standard screens and the monochrome display make it impossible to visualize complex cartographic images and imposes great limitations on cartographic scales of representation. Use of larger screens (30.5 cm and more), with colour display, provide a means of getting better and more complex images, but the price to pay for this improvement is a significant increase in required memory and in hardware costs. The memory needed for the creation and operation of an electronic atlas with HyperCard obviously depends on the number of stacks and cards used and the way in which data are entered on cards. The first micro-atlas presented in this chapter is contained on a single 800K diskette; the atlas on mines and minerals require three diskettes but it can be run on a hard disk. Sounds and images cost a lot of memory on a micro-atlas: In the Francophonie micro-atlas the 26 seconds segment of music at the beginning occupies 290K of memory.

It is evident that the operation of a micro-atlas created with HyperCard and containing more than 150 cards will require necessarily the use of a hard disk. In order to prepare a micro-atlas of 500 cards on a Macintosh computer, using HyperCard, gray scale images and sound, it is necessary to dispose of a disk space running from 8 to 10 megabytes.

Perspectives

In addition to the use of colour now permitted with SuperCard, one of the most interesting perspectives for the future development of electronic atlases on microcomputers is the coupling with analog or numerical videodisks. With its great storage capacity of images, fixed or animated, the videodisk can palliate the memory shortage mentioned above, with the result that the number of images to be inserted into the atlas can be multiplied drastically (up to 100,000 images on a single disk). Software like HyperCard is already able to manage a bank of images stored on a videodisk; these images can be displayed almost instantaneously, providing they have been coded appropriately and can be selected by using a keyword, a numerical code, a spatially referenced code or a mere pointer on a map displayed on a screen. The *Domesday Project* videodisk is probably the best example of that kind of achievement.

With such a technical capacity, the electronic atlas on microcomputer is destined to become a multimedia, multirelational and dynamic tool. The dynamic nature of geographic phenomena in space and time can be illustrated with animated time sequences, composed of filmed images or of computer graphics images transferred to a videodisk. The technical tools necessary to create such documents exist. It is the limited amount of reliable geographical data available in a computer form that undermines future development.

Conclusion

The rapid expansion of geographic information systems within the public and private sectors in recent years has come about through heavy investment. These systems have, in turn, contributed to the creation of a large amount of spatially referenced data and information. The investments can only be justified by the expected benefits accruing from better knowledge and management of the territories studied. But the information stored in spatially referenced data banks must not be reserved for the sole benefit of specialists in land management. After a suitable processing in order to facilitate its understanding by a non-professional public, this information must be made available to greater numbers of people. The creation of electronic micro-atlases, using data stored in GIS, constitutes one of the possible means to diffuse the information contained in these systems.

The advantage of micro-atlases which make use of hypertext concepts is to propose to the user a method for discovering spatial information that is similar to the way human mental process works. This includes the non-linear reading of geographic information, the setting of multiple relations between elements of information, the making of associations and correlations between data represented in different forms. In addition to its contribution to the diffusion of information contained in a GIS, the micro-atlas derived from a software such as HyperCard, can be turned into an instrument for the self-teaching of a non-specialized public. The micro-atlas can contribute to the development of the "informational" society that is rapidly supplanting the production society which we have known so far.

Acknowledgement

The micro-atlas *Mines et minéraux à la carte* was developed with the aid of a research grant from Energy, Mines and Resources Canada. We acknowledge the invaluable assistance of Professor Dean Louder (Department of Geography, Laval University) for revising the translation of this paper, and the contribution of Professor André Gamache (Department of Computer Science, Laval University) for giving us orientations on the hypertext literature.

References

Alberta Education (1979) *Junior Atlas of Alberta*, Alberta Education, Edmonton.

BBC Enterprises (1986) *The Domesday Project*, BBC Enterprises Limited, London.

Brousseau, Y. (1988) *Micro-Atlas de la Francophonie nord-américaine*, M.A. Thesis, Université Laval, Department of Geography, Québec.

Brousseau, Y. and M. Miller (1988) *La Francophonie nord-américaine à la carte*, Secrétariat permanent des peuples francophones, Québec.

Bush, V. (1945) "As we may think", *Atlantic Monthly*, July 1945, pp. 101–108.

Canada (1988) *Canadian Mineral Yearbook 1987*, Energy Mines and Resources Canada, Ottawa.

Canadian Mining Association (1987) *Mining in Canada. Facts and Figures*, Canadian Mining Association, Ottawa.

Caron, Lucie (1989) *Système hypertexte et son application aux données géographiques*, Université Laval, Department of Computer Science, Québec.

Carswell, R. J. B., G. J. A. de Leew and N. M. Waters (1987) "Atlases for schools: design principles and curriculum perspectives", *Cartographica*, Vol. 24, No. 1.

Cartier, M. (1988) *Hypercard. Premières analyses: présentation graphique et aspects médiatiques*, Communications Department, Université du Québec à Montréal, Montréal, unpublished.

Centre d'études en enseignement du Canada (1986) *L'interAtlas, les ressources du Québec et du Canada*, Centre éducatif et culturel, Montréal.

Englebart, D. C. (1963) "A conceptual framework for the augmentation of man's intellect", in *Vistas in Information Handling*, Vol. 1, Spartan Books, London.

Goodman, D. (1987) *The Complete Hypercard Handbook*, Bantam Books, New York.

Louder, D., C. Morissonneau and E. Waddell (1979) Du continent perdu à l'archipel retrouvé: le Québec et l'Amérique française, *Cahiers de géographie du Québec*, Vol. 23, No. 58, pp. 5–14. One full-page map: "Population de langue maternelle française à l'extérieur du Québec. 1970/1971".

Marcotte, L. and Y. Tessier (1987) "Applied research and instructional atlas design: the 'Ten Commandments' of l'InterAtlas", *Cartographica*, Vol. 24, No. 1, pp. 101–159.

Nelson, T. H. (1967) *Getting In and Out of our System, Information Retrieval: A Critical Review*, Thompson Books, Washington.

Nolette, C. (1986) *Thema, logiciel de cartographie statistique et thématique assistée par ordinateur*, M.A. Thesis, Université Laval, Department of Geography, Québec.

Openshaw, S. and H. Mounsey (1986) "Geographic information systems and the BBC's Domesday interactive videodisk", *Proceedings Auto-Carto London*, pp. 539–546.

Québec (1988) *L'industrie minérale du Québec 1987*, Ministère de l'Énergie et des Ressources, Québec.

Raveneau, J., C. Dufour and M. Miller (1989) *Mines et minéraux à la carte*, A micro-atlas distributed by the Department of Geography, Université Laval, Québec, Canada, G1K 7P4.

Raveneau, J., L. Marcotte and Y. Tessier (1987) "Le rôle du langage graphique dans le renouvellement de la conception d'un atlas pédagogique: le cas de L'InterAtlas", *The Canadian Surveyor/Le géomètre canadien*, Vol. 41, No. 3, pp. 313–339.

Shafer, D. (1988) *Hypertalk Programming*, Hayden Books, Indianapolis.

Shneiderman, B. and J. Morariu (1986) *The Interactive Encyclopedia System (TIES)*, Department of Computer Science, University of Maryland, College Park.

Siekierska, E. M. and S. Palko (1986) "Canada's electronic atlas", *Proceedings Auto-Carto London*, pp. 409–417.

Smith, R. M. (1987) "Electronic atlas of Arkansas: design and operational considerations", *Proceedings of the 13th International Cartographic Conference*, Morelia, Mexico, Vol. IV, pp. 159–167.

Tessier, Y. (1989) "Tendances récentes dans la production des atlas", *Cahiers de géographie du Québec*, Vol. 33, No. 88, pp. 73–88.

CHAPTER 11

Vehicle Navigation Systems

HINRICH CLAUSSEN

Institute of Cartography
University of Hanover
Federal Republic of Germany

Introduction

"Navigation" can be defined as the guidance of a vehicle from a starting point to a destination, and "positioning", a closely related discipline, will provide the means whereby the optimal route from a current location can be found. Both navigation and positioning are important aspects of vehicle navigation systems that will be considered in this chapter.

The methods applied for navigation have been available since people began to sail the oceans, but the equipment has improved over the centuries. In recent times developments in computer and communications technology have had a great influence on navigational methods, and this is particularly so for car navigation.

A decrease in the size and cost of computers and an increase in their speed has made development of vehicle navigation systems possible because, for vehicle navigation purposes, positioning and route findings have to be performed in real time with equipment which is compact in size. It is in this area that the development of microcomputer technology, as described in earlier chapters, is critical, as are developments in data capturing and data structuring.

As we have learned from shipping and aviation, maps are the most important navigational aid, and automated systems have come into existence largely because of progress in digital cartography.

The developments mentioned above correspond to a large demand for these systems which exists because of increasing volumes of traffic. A major goal in these circumstances is to enhance the management and safety of road traffic.

Concepts

Several automobile and automobile supplying industry leaders are currently engaged in the development of special electronic components for automobile navigation. These components provide vehicle navigation and step-by-step route guidance aimed at realizing a safer and more relaxed driving environment.

At present there are a wide range of systems emerging with various performance capabilities. These systems are either under development or already available on the market. They can be divided into two broad types, map matching and non map matching systems. The non map matching systems work with the support of analog map data only, while map matching systems use digital map data on a storage medium in the car as support for positioning and route guidance.

The minimum requirement for any system is knowledge of the position of the car at any time as well as the destination in the same coordinate system. This would enable the computation of direction and line-of-sight distance to the destination.

The method of positioning leads to another classification, because positioning can be performed either by support of infrastructure or autonomously, as shown in Fig. 11.1.

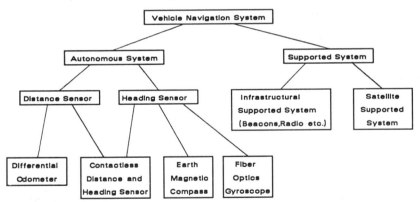

FIG. 11.1. Vehicle navigation systems.

For autonomous systems, the equipment needed for determination of the actual location is installed in the car.

Infrastructurally supported systems based on support of a satellite system, proximity beacons, or radio communications. These are necessary for the determination of the automobile's location. Proximity beacons can also be used for transmission of actual traffic data such as traffic jams or road works.

The advantage of having the most recent information about traffic events is offset by the considerable effort needed to provide the infrastructure, and by the disadvantage that the system has much less independence.

Autonomous systems adopt the dead reckoning method, well known from ship navigation and aeronautics. The principle is that vectors obtained by increments of distances and directions are added continuously and provide the actual position, as shown in Fig. 11.2.

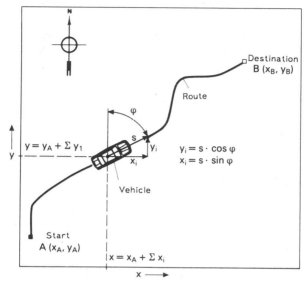

FIG. 11.2. Dead reckoning (Heintz and Knoll 1986).

Essential Sensors

Dead reckoning requires distance and heading sensors. Several distance sensors have been developed. As a source of signal for distance increments, pulse generators mounted on the non-driven wheels of a car can be used. They deliver a predefined number of pulses per rotation for the left and right wheels. Usually these sensors are used for cars with an antilock braking system. For positioning purposes they also have to manifest static behaviour because pulses that can be evaluated are needed at very low speeds (e.g. 1 km/hr). If the sensors are paired, recognition of the sense of rotation is possible, and position errors caused by reverse movement can be avoided.

The variation in direction can be derived from the difference between the distances covered by the inner and outer wheels in a curve as these values are proportional to each other, as is shown in Fig. 11.3. Errors may appear, especially if the radius of the wheels or the wheel track is not well known.

Earth magnetic field sensors (flux-gate magnetic compasses) measure an absolute heading to a natural fixed point (magnetic pole). However, the measurement with a flux-gate magnetic compass is prone to error which can be caused by local influences such as the starting of electric trams and magnetic variation caused by power lines and steel bridges.

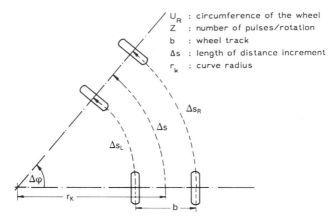

U_R : circumference of the wheel
Z : number of pulses/rotation
b : wheel track
Δs : length of distance increment
r_k : curve radius

(1) $\Delta s = r_K \cdot \Delta\varphi$

(2) $\Delta s_R = \dfrac{U_R}{Z} \cdot Z_R = (r_K + b/2) \cdot \Delta\varphi$

(3) $\Delta s_L = \dfrac{U_R}{Z} \cdot Z_L = (r_K - b/2) \cdot \Delta\varphi$

(2) − (3) $\dfrac{U_R}{Z}(z_R - z_L) = b \cdot \Delta\varphi \quad \rightarrow \Delta\varphi = \dfrac{U_R}{b \cdot Z} \cdot \Delta z$

(2) + (3) $\dfrac{U_R}{Z}(z_R + z_L) = 2 r_K \cdot \Delta\varphi \rightarrow \Delta s = \dfrac{U_R}{Z} \cdot z$

FIG. 11.3. Variation in direction (Neukirchner and Zechnall 1986).

Wheel sensors are insensitive to the short-term distortions which affect magnetic compasses. As a compromise, filtering and the addition of the interfering signals of both sensors provides improved direction values and increases the reliability of the car's location.

Non Map Matching Systems

Two European examples of non map matching systems are CARPILOT (Bosch-Blaupunkt) and CITYPILOT (VDO). These systems use an earth magnetic field sensor and a differential odometer. Examples of other systems are given by French (1987).

In both systems the driver has to enter the start and destination coordinates. In the VDO system, a light pen can be used to read the bar coded location of the start and destination on special maps. The on-board computer in turn calculates directions and line-of-sight distances to the driver's destination. On a simple liquid crystal display the computer presents the results of its calculations process. Utilizing either of these two systems will allow the driver to reach a specified destination within an error range of about 2–3% of the distance travelled.

Map Matching Systems

The positioning method described above is still affected by both systematic and random errors. If the sensors are properly calibrated most of the systematic errors can be avoided and the remaining errors can be regarded as random errors. Their size grows with the length of the journey. Position corrections can be made if a digital map data base is available on a mass storage medium in the car. By comparing a number of dead reckoned fixes with the topology and geometry of the map data, the actual position can be determined.

Systems Without Route Planning and Route Guidance

Systems without a route planning component require input of the preferred route and output of route information. To input the preferred route, a small digitizer can be used with which the selected route is digitized on a conventional paper map.

The navigation computer represents the route graphically on a display, and a fix symbol marks the location of the automobile. The locations are determined by dead reckoning and map matching, as described above.

A more advanced system of this class is offered by ETAK, which is available on the American market. This system continuously displays the location of the automobile and the surrounding road network on a CRT screen. The digital map data are stored on 3.5 Mbyte cassettes.

On the European market a product called TRAVELPILOT is available. It has the same performance as ETAK, but with a CD-ROM as the storage medium which has a capacity of 540 Mbytes. The destination can be entered in the form of towns and street names by means of a menu displayed on the CRT. Start and destination are then illustrated on the screen, and the display rotates to correspond with the heading of the automobile so that the road network is always observed from the driver's viewpoint. Different map scales can be shown according to the distance to be travelled and the detail required to be represented.

Systems With Route Planning and Route Guidance

A system of this type would be required to answer the following questions for navigation:

(1) What is the actual location of the automobile?
(2) What is the direction to the driver's destination?
(3) What is the line-of-sight distance to the driver's destination.
(4) What direction or road must the driver choose at the next decision point to reach his destination by an optimal route?

To recommend an optimal direction to a destination, route planning is required. Algorithms for this purpose are available in graph theory. Vehicle navigation systems have special requirements because routes have to be planned in real time. If the driver does not follow the recommended directions, the route planner has to offer a new optimal route. The algorithm of Ford (1956), Moore (1957), Domeschke (1972) and Pape (1980) provides routes in valuated graphs. This valuation stems from attaching the needed travel time to road sections. Route planning is performed in a hierarchical way: If the route is planned from the destination to all road sections, the result can be a complete table of routes leading to the destination. If the recommended route is not followed, the next optimal one can easily be identified, and the route guider can indicate a series of audio visual instructions from the planned route.

Prototypes of systems with this performance are EVA (Elektronischer Verkehrslotse für Autofahrer—Electronic Guide for Automobile Drivers), developed by the German company Bosch, and CARIN (Car Information and Navigation System) developed by Philips of The Netherlands. Both systems use synthetic voice generators to output driving instructions, for safety reasons. With the aid of a communications system, the driver enters the destination for the trip. The actual location of the vehicle is constantly known, even while the vehicle is parked. The location and route planning system provides input to the navigation system, and the on-board computer processes the input data, calculates an optimal route and prepares step-by-step route guidance. The communications system presents this information to the driver by synthetic voice and by visual display.

Infrastructure Assisted Systems

While this approach is no longer in use in the United States (French 1987), further developments have been undertaken in Europe and Japan. There is a demand for these systems where traffic situations change frequently because they allow traffic-dependent vehicle guidance.

In West Germany the ALI System (Autofahrer Leit- und Informationssystem—Automobile Driver Guidance and Information System) was proposed in 1970. Subsequently a field test was prepared and carried out during the years 1979–82 on an autobahn test network of about 150 km in length. Communications between the car and a central control computer were performed by means of induction loops. Information such as speed, type of car, desired destination, etc., were transmitted to the computer where the actual traffic situation was derived and the individually recommended route was then calculated. Driving instructions were then sent to the car and given to the driver in visual form whenever an induction loop was passed at an intersection or highway exit. In addition, warnings about road conditions such as traffic jams, ice, fog, etc., were transmitted to the car and displayed on the vehicle unit. This field test was successful and provided the first approach towards an individual guidance system.

Further development was evoked by the decreasing costs of semiconductors as it became possible to have more computer intelligence in an on-board computer. The ALI-SCOUT system consists of infrastructure and in-car equipment, and is a result of an evaluation of existing navigation and guidance systems. Infrared beacons, which are installed at important intersections of the road network, are used for the transmission of information. The actual traffic situation can be derived from travel times sent to a central traffic computer. Traffic dependent driving instructions are then calculated and sent to beacon control units. These instructions are received by the on-board computer in the car and given to the driver visually or acoustically when passing a beacon. The vehicle's position is obtained by dead reckoning and map matching methods.

The ALI-SCOUT system is now being tested in a large scale field test known as LISB (Leit- und Informationssystem Berlin—Guidance and Information System Berlin). This test covers the entire street network of West Berlin. Two hundred and twenty intersections are equipped with infrared beacons which communicate with 700 test drivers.

Auto Guide, a system similar to ALI-SCOUT, is being developed by the Automobile Association (AA) in Great Britain.

Digital Map Requirements

All functions of a navigation system make their own particular demands on the on-board digital map. For instance, the map matching function requires geometric information with sufficient accuracy about the road network. This information is also needed for the calibration of vehicle parameters.

A further important function of a navigation system is address conversion. In most countries the usual way to specify addresses is in terms of town names, street names and house numbers. The system, however, uses internal addresses specified by, for example, identification numbers of roads or other objects. Uniqueness of addresses plays a significant role. To guarantee uniqueness of a street name outside the context of a particular municipality, municipality names as well as higher order administrative areas are required. The same holds true for objects such as restaurants, cinemas, petrol stations, etc., assuming they are recorded. The knowledge of house numbers enables the specification of a more precise location on a street.

Route planning is possible if the topology of the road network is known, i.e. if roads are directly connected at an intersection or if they cross each other by means of a bridge or viaduct. The optimal path can then be formed by connected road sections. The performance of the route planning function can be enhanced with help of a road classification that leads to a valuated graph. Travel time can then be a criterion for the optimal route. If regions are just to be passed through, then roads of lesser priority can be ignored in order to save computation time. In this case the detailed structure of intersections is not important, but these constructions can be aggregated to one composite object, as shown in Fig. 11.4. As well, knowledge about

 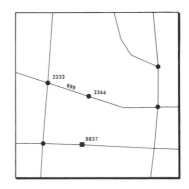

FIG. 11.4. Simple and aggregated objects.

traffic restrictions and prohibitions such as one way roads, prohibited turns, etc., should be taken into account to avoid planning illegal routes, and dimensional restrictions are particularly useful for truck drivers (i.e. height, width, length).

For clear driving instructions the route guiding functions need an aggregated representation of the road network: Road sections form roads; internal links and junctions form intersections, etc. Special features like roundabouts require special instructions and, for this reason, the route guider must know the road elements belonging to a roundabout. Furthermore, the information represented on signposts is useful because guiding instructions may then refer to it.

To improve the readability of a map image produced by the position display function, the map should also contain a number of land use units, railways, rivers, canals and other landmarks.

Data Model and Data Content

The main data sources of digital map data for car navigation systems are existing databases created for different purposes, or analog road maps. Acquisition of road data for vehicle navigation systems is very costly and time consuming. Both time and cost can be decreased if this work is done only once and in a way that allows the data to be accessed by several systems. This requires standardized methods for acquisition and exchange of data. In the United States large databases exist at the Bureau of Census. These data have been adapted to car navigation demands made by ETAK and the data sets have been extended accordingly. In Europe a project called DEMETER (Digital Electronic Mapping of European Territory) has been established. Its goal is to create standards for digital map data which will primarily be used for car navigation purposes. The first result of the work on this project is a standard named GDF 1.0 (Geographic Data File Release 1.0) which consists of a Specification of Data Content (SDC), a Specification for Data Acquisition (SDA) and an Exchange Format (EF).

The data model of GDF is feature oriented. A feature is a formalized entity made to represent a topographical object. A group of features that is strongly related is called a layer, and the most important for car navigation is the layer entitled road. There are many other layers, e.g. administrative areas, buildings, railways, waterways, etc. Features have particular quantitative and qualitative properties that are described by means of attributes. Interrelationships can be recorded explicitly, while shape and location of features are expressed by means of geometric data.

Most important for some purposes, however, is that data are topologically structured because topology enables the description of connections between neighbouring features (Peuquet, Chapter 4). Finding a path through a network is a topological calculation. For map matching purposes the topological property of connectedness is required. Topologically the data are stored as nodes (0-cells) and chains (1-cells). Connections are indicated by start and end nodes of chains or by a pointer from nodes to connected chains. Nodes are located at all intersections of chains. On the object level connections of road sections are explicitly indicated.

Locations are given by pairs of coordinates that can be attached to nodes and chains. In the case of chains they describe the shape by means of intermediate points. By referring from features to topology, geometry can be attached to features and geometry of punctuated features, such as landmarks, can be directly attached.

Essential for a road data base are attributes, and important attributes for road are road class, one-way flow and turn restrictions, and meanings of traffic signs. Administrative areas may have area codes and type names. For buildings and services attributes like brand names and opening hours may increase serviceability. Some of them can be captured directly from annotations and symbols on paper maps, but many, such as content of traffic signs and house numbers, require field work. After the survey they must be attached to those data gathered from maps, and this is performed by pointers from features to attribute identifications. Proper names, like street names and names of administrative areas, are handled separately in the same way.

Apart from relationships between different data types like attributes and features, relationships between different feature types can be indicated explicitly. These relationships, where more than one feature is involved, concern prohibited turns, rights-of-way and signpost information. Other relationships are created to give the possibility of describing situations in which the logical tie between features is not equivalent to their spatial one, e.g. a building has its entrance on a road other than the one which is nearest. A further type of relationship is created to describe which road is the upper and which the lower at level crossings.

Besides spatial information, a data set has to contain global information that is needed for correct interpretation of data sets. These data are also capable of keeping the external documentation to a minimum. Each data set is subdivided into sections and layers, whereby a section is a spatial subdivision and a layer a logical subdivision. The data set itself and the subsets have to be preceded by a corresponding header which identifies and briefly documents the data collection in question. Also important is the group of bibliographic data which explain the lineage by means of identification of source documents. A directory describes the content, size and division into subsets of a data set. Geometric data do not have to be provided according to a unique reference system, but it is essential that geodetic data are indicated. These have to provide information about items such as horizontal datum, projection systems, projection parameters, origin of the coordinate system, coordinate off-sets, vertical datum, geoid ondulation and the magnetic declination. The quality of a data set should be given in terms of geometrical accuracy, the time of data collection, completeness of features and attributes, and error rates in attribute values.

According to the conceptual data model, an exchange format has been created within DEMETER. The goal of the exchange format was to meet the following requirements (Claussen et al. 1989):

— It should have enough representation power to cover the special needs of navigation systems.
— It should have the possibility to represent topologically structured data.
— It should allow independence between the representation of the semantic, topological and geometric information of a particular feature.
— It should be open ended so that it can be extended with new fields, records, feature codes and attribute types, without the need to redesign the format completely.

Outlook for the Future

The advanced technology for vehicle navigation systems has already been developed, and some very promising tests have been performed. In limited urban areas infrastructure supported systems may provide an appropriate method for fleet management while for individual guidance purposes autonomous systems are probably more appropriate. The prerequisite is that digital road data can be provided for extended areas and, even more important, are the traffic-relevant attributes which can only be captured by field survey. This requires a major effort and is only possible by support from all industrial and governmental agencies which are interested in vehicle navigation. The databases have to be updated continuously and, in addition, actual traffic data have to be provided. In this area the Radio Data Signal (RDS) is very promising. The possibility of communication from infrastructure to vehicles will be a first step towards enhancing the benefits of the systems described. However, vehicle to infrastructure data communication could provide destination information to central traffic management systems for planning optimal traffic flow. Finally, for wide propagation of car navigation systems, standardization is required in the fields of navigation, digital maps and mobile data communication.

References

Claussen, H. (1989) "GDF—Ein Austauschformal für geographische Daten", *Nachrichten aus dem Karten und Vermessungswesen*. Reihe 1, Heft Nr. 103, pp. 37–44, Frankfurt am Main.

Claussen, H., L. Heres, W. Lichtner and D. Schlögl (1988) "Digital electronic mapping of European territory (DEMETER)", *Proceedings 18th International Symposium on Automotive Technology and Automation*, Florence.

Claussen, H., L. Heres, P. Lahaije, W. Lichtner and J. Siebold (1989) "GDF, A proposed standard for digital road maps in car navigation systems", VNIS '89 Conference, Toronto.

Domeschke, W. (1972) "Kürzeste Wege in Graphen: Algorithmen, Verfahrensvergleiche", *Mathematical Systems in Economics*, Heft 2, Meisenheim.

Ford, L. R. and D. R. Fulkerson (1956) "Maximal flow through a network", *Canadian Journal of Mathematics*, No. 8, pp. 399–404.

French, R. L. (1987) "Automobile navigation in the past, present and future", *Proceedings Auto-Carto 8*, pp. 542–551, Baltimore.

Heintz, F. and M. Knoll (1986) "Kraftfahrzeuginformation: Ergonomic und Technik", *Bosch-Technische Berichte*, Band 8, Heft 1/2, pp. 32–46, Stuttgart.

Lahaije, P. D. M. and R. H. H. Wester (1988) "Efficient road-map management for a car navigation system", *Philips Journal of Research*, Vol. 43, No. 5/6.

Lichtner, W. (1988) "The integration of digital maps in vehicle navigation systems", *Proceedings of the 13th International Cartographic Conference*, Vol. IV, pp. 1–10, Mexico.

Mark, D. M. (1987) "On giving and receiving directions: cartographic and cognitive issues", *Proceedings Auto-Carto 8*, pp. 562–571, Baltimore.

Moore, E. F. (1957) "The shortest path through a maze", *Proceedings of the International Symposium on the Theory of Switching*, April, Part II, pp. 285–292.

Neukirchner, E. and W. Zechnall (1986) "EVA—Ein autarkes Ortungs-und Navigationssystem für Landfahrzeuge", *Bosch-Technische Berichte*, Band 8, Heft 1/2, pp. 7–14, Stuttgart.

Neukirchner, E. P., W. Suchowerskyj and W. Zechnall (1988) "Vehicle navigation system status and developments in Europe: an overview", *Proceedings 18th International Symposium on Automotive Technology and Automation*, Florence.

Pape, U. (1980) "ALGORITHM 562, Shortest path lengths", *ACM Transactions on Mathematical Software 6*, No. 3, pp. 450–455.

Siebold, J. (1989) "Nutzung digitaler Strassendaten für Kfz-Navigationssysteme", *Nachrichten aus dem Karten und Vermessungswesen*, Reihe 1, Heft Nr. 103, pp. 113–120, Frankfurt am Main.

Thoone, M. L. G. (1987) "CARIN, a car information and navigation system", *Philips Technical Review*, Vol. 43, No. 11/12.

White, M. (1987) "Digital map requirements of vehicle navigation", *Proceedings Auto-Carto 8*, pp. 552–561, Baltimore.

CHAPTER 12

Cartographers and Microcomputers

JEAN-PHILIPPE GRELOT

Institut Géographique National
136, bis rue de Grenelle
75700 Paris, France

To conclude this book on microcomputers in modern cartography, it is useful to complement the papers on the technical aspects by considering the precise role of cartographers in the effervescence of the disciplines dealing with information.

The changes designated by the term "information revolution" are not specific to cartography; they affect all of society and all of its actors whether they be decision makers, producers or users. In a society where information overload is increasing the most efficient means of communication will be favoured, and here is where cartography has a major role to play. It is not sufficient just to make such a declaration: on the contrary, a great deal remains to be done for, as Taylor has written, "As an applied, visual discipline with communication as a central element, cartography has much to offer in the information era, but if that potential is to be realized then considerable re-thinking will have to take place and new enthusiasm generated in response to the challenges facing us" (Taylor 1988, p. 1). He further develops this theme in the opening chapter of this volume.

Cartography has been styled as an "emerging discipline" (Wolter 1975), whilst at the same time cartographers are aware of a hidden threat which sees cartography being diluted to become merely an appendage to other disciplines, as it was for a long time a necessary yet minor part of geography. The choice between the emergence or, alternatively, the increasing subordination to other disciplines will not be decided in cartography's favour unless, forgetting for a moment our maps or our more recent display consoles, we estimate the economic and social stakes and assimilate them in our actions. The strength of cartography is not in the number of scientific conferences, or the number of communications or participants, but in the discipline's ability to meet both the explicit and implicit expectations of contemporary society.

In this process the microcomputer is a powerful tool, and this chapter will deal with its implications for cartography, including both the economic and technical aspects and the impact on the importance and transmission of knowledge.

Technical Background

Of the technical changes which have affected our profession during the last decade, and described in several chapters in this volume, two seem to be of particular importance for consideration by cartographers. They are satellite and data processing technologies. More precisely, because both these technologies are several decades old, one must refer here to the characteristic that both have acquired over the last decade, which I will name "democratization".

During the last decade, Earth observation satellite imagery and data processing emerged from the laboratories and specialized agencies. The desk top and the home computer are already the working tools of cartography as such, if not more so than the pen and the scriber. Why then are these changes so important?

With Earth observation satellites we can discover the entire world from our offices. It is no longer necessary to go into the field; it is no longer necessary to send photographic aircraft as satellites record their images continuously with frequencies which generally cannot be equalled by the time required in sending teams to distant places. Although the image resolution is still well below that of aerial photographs, and the stereoscopic quality or the probability of obtaining pairs of stereoscope images must still be improved, considerable progress is now being made in these areas. With image analysis techniques based on automatic correlation, and by adding a good dose of expert systems and a pinch of artificial intelligence, the constitution of medium or small scale data bases (from 1 : 50,000) will, in five or ten years' time, be entirely automatic. It will be a change as significant as the transition from direct topographic surveys to photogrammetry. The state of world cartographic coverage and its rate of revision (Brandenberger Ghose 1989) is poor, but with new techniques perhaps we shall at last have the possibility of improving the situation.

The most extraordinary fact is that required digital processing can be undertaken on a workstation, which will be a normal tool for cartographic technicians. Furthermore, it is because these workstations are widely distributed that one can envisage processing the considerable mass of recorded data. After three years' operation, the SPOT satellite has already recorded 900,000 images, each covering 3600 km². Even if one reduced this number to 25% of images with acceptable cloud coverage, there remains 5000 to 7000 gigabytes of data for our information.

As far as computers are concerned, few technical fields have evolved at a similar pace. Electronic chips are now used everywhere; from aircraft to washing machines, from automobiles to cameras. By reaching increasing strata of our population and by becoming an object and a learning tool in schools, the microcomputer has a multiplier effect. It supplies results more rapidly than other methods, and the earlier availability of the results makes it possible, in turn, to accelerate the process which required them. A typical example is the internationalization of financial markets where transactions are displaced in accordance with the rhythm of the opening of the Tokyo, London and New York Stock Exchanges. The transactions are transferred, but those who negotiate are elsewhere and the investors for whom they act are themselves fixed in space. This is only possible because the data or data banks are universally accessible to networks based on simple microcomputers.

"Democratization" is the term chosen to describe the characteristic acquired by satellite imagery and data processing over the last decade. Democracy, according to the Larousse Dictionary, is a governmental system where the people exercise the sovereignty and, by extension, to "democratize" means putting something within the reach of all classes in a society. As a result of low selling prices and the efforts devoted by certain countries to education, microcomputers are now within the reach of a large number of individuals. This wide distribution does not itself create access to sovereignty by the people, but it induces undeniable changes in behaviour, which is a requirement for accessibility to and availability of information at the same time as new exchange circuits are set up. In this respect one speaks more and more of networks and the roles they play in society whether they concern collective economic networks or individual relations networks. At the same time the transition from the mainframe computer to the microcomputer is from a collective use (the entire group of users of a computer) to an individual use, and in that process the location of knowledge is necessarily displaced.

With the miniaturization of computers the number of uses has literally exploded, and from general information systems to very specific dedicated applications, almost anything seems possible. For example, consider the evolution of aid to the motorist (see chapter by Claussen). The first example in this field is assistance in selecting a route. For that operation the data base describes the logical structure of the road network whose elements are subdivided into sections designated by towns and cross roads. For each section and length, the average travel time and the travel time linked with traffic conditions such as dense traffic, rain or ice, is indicated. The software includes a user interface for interrogation and a display or a plot of the result, together with an operational research algorithm, the objective of which is to calculate the shortest route under constraint (minimize the distance travelled or minimize the travel time). This program can be implemented on a special microcomputer or distributed from a central data base.

The second example concerns aid to automobile navigation. The system installed in a vehicle can be independent, or dependent on the surroundings for receiving information in real time. In the independent case, the automobile is provided with a data processing system. The database includes a logical and geometric description of the road network (because it is necessary in this particular case to spatially reference the vehicle in the network); sensors supply data on the absolute or relative position (with respect to the departure point or reference points); and a computer compares the sensor data with the cartographic database to determine the vehicle position. In its primitive version the system thus supplies spatially referenced information. This is a very specialized function, and one can optimize the system by constructing a processor or specialized architecture.

The spatial reference system can be coupled to the route selection system discussed above. One then enters the field of intelligent systems for assisting automobile drivers, where one can go as far as real time guidance by the continuous sending of information, making it possible to recalculate the optimal route depending on the instantaneous analysis of traffic conditions.

These three examples, route selection, navigation and driving aids, are dedicated systems. The user supplies a few parameters (destination and, when necessary, departure point), and the system provides the result automatically without any other action.

The final objective is to be able to install individual equipment in each vehicle. The development costs are such that only mass consumption will make it possible to attain unit prices which will make the equipment attractive.

The experience acquired by geographic information systems' experts now leads to advocating pragmatic and progressive steps. As the technology is no longer a barrier to the establishment of new systems, it is now necessary to concentrate more on its use.

The approach is comprehensive and links a conceptual phase (definition of the information required for the functioning of the system), an organizational phase (definition and construction of the collection network and information updating), and finally, and only then, a technical phase where one installs the equipment and software (Service Technique de l'Urbanisme 1989, p. 27). The technology, as already stated, no longer imposes insurmountable constraints. As core memory capacity and mass storage capacity, based on optical disks, increases along with an increase in processing speeds, the telecommunications barriers fall progressively. Through the use of cables or satellite links the computers, both small and large, communicate between themselves. Information is shared on a worldwide basis and, equally important, it is shared instantaneously.

This immediacy modifies relations between the reader and the map. The instantaneity of the display on a screen reduces the discovery time and the mental preparation time of the reader. The modification is similar to that of a traveller: Transatlantic liners provide crossing times for detachment from the departure zone and for preparing oneself for new territories; the New World for the Europeans, or their historical and cultural roots for the Americans. The aeroplane does not shorten distances but it does shorten time; it compresses the time for dreaming, imagination and detachment which are sources of creativity due to one's liberation from stereotyped models. In the aggressive instantaneity of the display, aesthetic seduction disappears and gives way to the cold world of numbers.

Image-less Maps

Actually it is a new form of cartographic communication that is being proposed. This creates an important problem because microcomputers take tools away from a small group of specialists, the "auto cartographers", and put them within the reach of vast layers of the population including not only individual users (the general public), but also decision makers whose actions have consequences for all the communities involved.

Our knowledge has progressed considerably in recent years due to studies of information theory and the design of communications models. They make use of visual perception and have led to the construction of a mental model based on the pre-existing knowledge of the reader. Maps, considered as images, are an intermediary in that process, and "signifiers" an expression of the "signified". Going from maps to databases then has several consequences.

The image displayed on a screen is transient as opposed to the paper map which is permanent. It is transient, but one can play with the display sequence of its elements; for example, on a general background displayed quasi-instantaneously one returns to plot slowly, with a specific colour, the road to be followed for linking two localities. The image can be composed or recomposed by the user. On a population density map colour sequence can be selected which risks violating the rules of graphic semiology. However, quite often the display will have no legend due to the size of the screen or the scale will not be indicated, which is not critical in most cases but becomes so if a given use requires a geometric accuracy greater than that supplied by the data. The user will not always have the opportunity to distinguish between admissible enlargements (those supported by the geometric accuracy of the data), and ergonomic enlargements whose only advantage is that of ease of reading.

It is necessary, in a way, to introduce a new cartographic culture where one is interested less in images and more in cartographic or geographic objects.

To change the intermediary by passing from the image map to the database map must also be accompanied by a transfer of the mediation: the conceptual model of a map and that of a database are different once the data are not only graphic but cartographic as well. These models do not represent exactly the same thing, without which the information would only provide a low added value.

For example, the topologic relations between objects on a map (intersection, adjacency, inclusion, exclusion) are implicit. Visual perception uses them because the reader's eye moves from one spot to another on the map. However, the order of passing from one spot to another is not determined in advance, which gives freedom in the reading of any image. On the other hand in a database it was fairly quickly recognized that it was necessary to use the data structures which represent the topologic relations. By moving from an implicit document to an explicit document one does not simply change the image, one also modifies the relationship between the reader and the object being viewed. The topologic relations which have become explicit are going to be used by the processing system or the display system, making it possible to show something else or show it in a different way. In the process the reader has lost part of his reading freedom.

Less aesthetic seduction, less reading freedom: what is the price of these lost pleasures?

Knowledge Recomposition

The above question also represents a displacement of the centre of interest of the cartographic community. One can measure the path covered by remembering that the International Cartographic Association was set up some thirty years ago because of concern arising from the technical evolution of the graphic arts (ICA 1987, p. 9). Today our discussions are centred on geographic information systems. Even more than the modification of the knowledge required by a cartographer, the change through which we are living is characterized by a phenomenon for which we are unable to envisage the consequences. This concerns the transfer of an important part of technical knowledge into a machine which modifies profoundly the relations between participants whom, in order to simplify matters, we will limit to the information collector, the cartographer/producer and the reader/user. In fact data processing systems pass progressively from the simple program to the expert system by using the resources of what is usually called artificial intelligence. Deliberately one seeks to give Boeotian users the processing tools which implicitly or transparently integrate the basic theoretical knowledge concerning cartography.

This being so, one is participating in a recomposition of knowledge which profoundly affects the cartographic profession. The cartographic processing chain includes two basic links: The cartographer who has transformed a prototype into a map, and the reader who has read the map and made a mental model of it but who interfered very little with the plotted data. Today the cartographic function, the first link in the chain, is divided in two: on one hand there is the acquisition, i.e. the transformation of a prototype into a database and, on the other hand, the graphic production which, in addition, becomes optional, consisting of processing the database in order to produce a drawing. The manufacturers of computer equipment and, even more, those producing computer software do their utmost to analyse the process used in great detail in order to put them in black boxes at the direct disposal of users, which carries them towards the second link in the chain.

Take a simple example. Any cartographer worthy of the name has spent several hours of his career drawing latitude and longitude grids which have specific advantages. This is more than a mathematical recreation or accuracy exercise, it is training in communication. As the world is not flat, any plane representation is a deformation which has to be mastered. Today any microcomputer can be used to plot over thirty of the most typical projection systems without the user needing to know the mathematical properties of equal area projections, conformal projections or Tissot ellipses.

Likewise, for data acquisition the problems of coordinate transformation no longer arise and it is sufficient to have available field measurement equipment and follow an operational procedure. The raw data are directly recorded by the equipment then transferred to a microcomputer which calculates the geographic coordinates or plane coordinates in any system whatsoever.

It would be useless to philosophize for a long time as to whether these changes are good or bad. They exist. It is in the order of things that technology evolves, and the landscape of professions evolves with it. Darwin's theory applies not only to biological species, and *Homo sapiens cartographicus* must accept being a mutant thereof.

What is our place in the data processing melting plot?

What one observes with the spread of microcomputer data processing is the standardization of the specific techniques of production and also of data processing itself. It is no longer necessary to be an eminent statistician to be able to make an ascending hierarchical classification or calculate a regression surface. Soon it will no longer be necessary to be a geodesy expert to calculate a triangulation: a GPS receiver, a microcomputer and a technical manual will be sufficient in many instances. "Know how to use" will replace "know".

Beyond the few experts from whom one will draw the rules to feed into the systems will one observe a dilution of cartographic knowledge? What place will there be for cartographers, who will be drawn from a wide variety of horizons, between high level experts and users? Is one moving towards a group of general practitioners in geographic information: on the one hand those who know the description and modelling techniques for geographic objects, and thematic specialists for each utilization field who are capable of developing and using applicative software?

Whatever happens, the pressure of the demand is fundamental, and systems will develop with or without the participation of cartographers. In order to decide on the place to be occupied and the path to be followed, it must be remembered that our current data processing systems were not produced for cartographic investigations; they result from computer assisted plotting systems, image processing systems and graphic industries systems (see chapter by Coll). What one demands from technicians today is that they be capable of assembling the pieces of a puzzle, not for producing an attractive product or a technical feat but to meet precisely a requirement coming from the exterior. "Today we have a real need for 'engineers of the third type', having a truly technical and scientific culture and knowing how to design and manage increasingly complex production systems, along with all the human and economic aspects which overlap with them. For a long time trained as a designer or rather as a calculator of objects and well defined procedures, the high level engineer must, henceforth, above all, manage a multitude of interfaces between varied sub-systems, organizations and techniques. He must know how to master the flux of events rather than domineer a world of status objects. He can no longer ignore the upstream (research) nor despise the downstream (market). The diversity of competences, the double opening on the varied engineering sciences and on the human sciences are a basic factor of modernity" (Veltz 1989, pp. 70–73).

In short, that which technological evolution requires of a cartographer is always to be more at the heart of the world and its mutations. As cartographers are basically spectators of the world that should not give rise to insurmountable problems. It is certain that technique is not an end in itself, and the influence of the economy has a greater impact, which is good, because there is a demand from the market for widespread installation of automatic cartography systems.

Market Context

The digital cartography market benefits from a combination of favourable circumstances. Technical maturity has been or soon will be attained, as was argued at the beginning of this chapter. A major part of cartographic knowledge is being integrated into systems that are henceforth directly usable by technicians with very little cartographic training. To that must be added two economic aspects, the challenge of time and the challenge of costs.

Cartography is information processing. The more rapid that processing becomes the more decisions will be made in deliverably shortened times. That will be the determining factor in certain instances such as military combat and natural or industrial disasters. Less dramatically, the selection of a commercial location will dictate the future of an enterprise with respect to its competitors. In order that these decisions can be made the information must not only exist, it must also be known to exist. The time compression phenomenon, or the reduction of the time allowed for decision making, will affect the time required for information construction as well as the time required for its extraction.

With microcomputer information processing, one amplifies still further a digital geographic information paradox of which we have progressively become aware. There is a strong correlation of the cost and delivery time economic characteristics between, on the one hand, the processing hardware (the computer and, to a lesser extent, the software) and, on the other hand, the data. One can quickly become accustomed to judging the operational character of an application with a prototype without worrying about the feasibility of extrapolating to a significant territorial scale which woud be necessary in order to know the cost of data acquisition, what time duration will be required and how data revision will be undertaken. In other words, there is an imbalance between the cost of acquiring the data and the cost of accessing that data. Let us examine these two aspects in more detail.

When a decision is made to invest in a geographic information system the investment is easily calculated. This includes an investigation leading to a precise definition of the system, preparation of a prototype ready for validation, together with the hardware and software. Sometimes it is necessary to add the cost of rearranging technical facilities. The cost of personnel training must always be added, together with the cost of interfaces, including telecommunications systems. However, that is only the tip of the iceberg. In fact, data acquisition costs are frequently of a greater magnitude than the direct investment costs. For an urban database designed for the operational management of utility distribution networks, costs are divided in a ratio of 10% to 15% for the investment and 90% to 85% for the data acquisition. This disproportion becomes much more marked as the data processing hardware makes more use of microcomputers and as technical evolution necessitates. However, buyers are usually more concerned with immediate expenditures and investment costs rather than with deferred expenditures and the composition of a database. Even so, no mention has yet been made of data revision which could amplify the problem even more.

It is for these reasons that economic considerations of processing a datum have to be taken into account. This cost combines the depreciation of all the investments, the system's operational costs, the data acquisition costs and the costs of its revision over the years.

One compares that value to the cost of data access, which is currently quite often merely a simple copy of the database on magnetic tape or floppy disk and may in future be by teletransmission or on a CD-ROM. That cost, beneath commercial tariffing for the right to use the data, is always very small with respect to the cost of possessing the data as defined above. For a very modest cost the data acquirer can thus duplicate the product he has bought and distribute it widely without the knowledge of or at the expense of the initial author. It is extremely difficult to protect oneself against that kind of abuse.

The situation is one of low cost and great ease of access to the data and high cost for setting up the data bank. The user of the system is not always aware of the financial aspects, which are the responsibility of the decision maker. On the other hand the user, whom we have seen is not necessarily knowledgeable about cartography, will always be sensitive about delivery times, and data acquisition can prove to be a long process. However, in order for a geographic information system to become really usable, the data available should cover almost all of the area in question (Bie 1985). Otherwise, and according to Murphy's Law, queries have a greater probability of relating to missing data than the proportion of data which remains to be acquired. In order to avoid this problem, one is led to construct a progressive application where the thematic aspect has a considerable advantage over the geographic extension. Instead of working region by region to acquire all the data on an area and to offer in total all the operations, certain operations for which the data can be fairly rapidly acquired over the entire area are chosen. One accustoms the user to the system and progressively increases the application domains. The geographic information system designer must thus undertake some engineering.

We have just seen that one is obliged to offer some applications rapidly. That is the expression, in the domain of the installation of geographic information systems, of a general characteristic of market evolution: one passes nowadays from a supply market to a demand market. In a supply market, which corresponds to an economy of scarcity, the producer defines the products, the market absorbs them, and there is always a buyer. Maps printed on paper are supply driven: the producer, after having analyzed requirements, defines a product aimed at satisfying a common subgroup of requirements, and the buyer must adapt himself to the product. In a demand market one must, on the contrary, adapt the product to each client. It is evident that cartographic databases better satisfy that constraint than printed maps. To the producers or production organizers who dominate in a supply market, it is now necessary to add sales staff and marketing experts whose roles are to continuously ensure the adequacy of the product to meet needs.

Conclusions

My arguments have sometimes gone beyond the strict framework of the cartographer and the microcomputer. However, the evolution of a profession is long, whilst that which today requires major data processing means will function tomorrow with a microcomputer.

Two basic tendencies can be distinguished in this chapter. The first concerns cartographers themselves. It is necessary to prepare for a rearrangement of the profession with, on one hand, high level knowledge engineers who, to a certain extent, will the geographic information theoreticians, and, on the other hand, application technicians whose skills will no longer be primarily manual dexterity but instead the capacity to use workstations with increasingly rapid performances. In that landscape "foreign bodies" coming from the data processing industry will seek to take a place. Let us endeavour to cooperate with them and hope they will feel the need to consult with us, otherwise they will do it without us and there will only remain our cartographic conferences where we can pity ourselves.

The second basic tendency concerns products and, increasingly, the services required of us. We have seen the importance assumed by information and networks in all their forms. In other words, one is no longer interested simply in existing structures. The cartography of exchanges, movements, the networks themselves, the cartography of evolutions is going to be sought from us.

With microcomputers, technical obstacles will disappear and the demand will increase. The field is favourable to us and will only require a small effort on our part to make known our competences and have them recognized. That is the image we must give through our professional associations, our conferences and our publications.

References

Bie, S. (1985) "Database technology: the fundamental issues for the mapping industry", CERCO Intensive Course in Computer Assisted Cartography.

Brandenberger, A. J. and S. K. Ghosh (1989) "Status of the world's topographic and cadastral mapping", 4th Regional Cartographic Conference of the United Nations for the Americas.

ICA (1987) "Twenty-five years", International Cartographic Association, compiled by F. J. Ormeling, Sr.

Service Technique de l'Urbanisme (1989) Informations géographiques: des inventaires aux systèmes?

Taylor, D. R. F. (1988) "Direction for an emerging discipline", International Cartographic Association, Newsletter No. 12.

Veltz, P. (1989) "Enseignement supérieur et recherche technologique: quelques réflexions générales", PCM –Le Pont No. 3.

Wolter, J. (1975) "Cartography, an emerging discipline", The Canadian Cartographer, Vol. 22, No. 4, pp. 19–37.

Index